New Insights Into the Stratigraphic Setting of Paleozoic to Miocene Deposits - Case Studies from the Persian Gulf, Peninsular Malaysia and South-eastern Pyrenees

Edited by Gemma Aiello

Published in London, United Kingdom

IntechOpen

Supporting open minds since 2005

New Insights Into the Stratigraphic Setting of Paleozoic to Miocene Deposits – Case Studies from the Persian Gulf, Peninsular Malaysia and South-eastern Pyrenees
http://dx.doi.org/10.5772/intechopen.75278
Edited by Gemma Aiello

Contributors
Haylay Tsegab Gebretsadik, Chow Weng Sum, Valenti Turu, Hadi Sajadi, Roya Fanati Rashidi, Gemma Aiello

Notice
Statements and opinions expressed in the chapters are these of the individual contributors and not necessarily those of the editors or publisher. No responsibility is accepted for the accuracy of information contained in the published chapters. The publisher assumes no responsibility for any damage or injury to persons or property arising out of the use of any materials, instructions, methods or ideas contained in the book.

First published in London, United Kingdom, 2019 by IntechOpen
IntechOpen is the global imprint of INTECHOPEN LIMITED, registered in England and Wales, registration number: 11086078, The Shard, 25th floor, 32 London Bridge Street
London, SE19SG – United Kingdom
Printed in Croatia

British Library Cataloguing-in-Publication Data
A catalogue record for this book is available from the British Library

Additional hard copies can be obtained from orders@intechopen.com

New Insights Into the Stratigraphic Setting of Paleozoic to Miocene Deposits – Case Studies from the Persian Gulf, Peninsular Malaysia and South-eastern Pyrenees
Edited by Gemma Aiello
p. cm.
Print ISBN 978-1-83880-443-5
Online ISBN 978-1-83880-444-2
eBook (PDF) ISBN 978-1-83880-434-3

We are IntechOpen,
the world's leading publisher of
Open Access books
Built by scientists, for scientists

4,200+

Open access books available

116,000+

International authors and editors

125M+

Downloads

Our authors are among the

151

Countries delivered to

Top 1%

most cited scientists

12.2%

Contributors from top 500 universities

CLARIVATE ANALYTICS
BOOK CITATION INDEX
INDEXED
WEB OF SCIENCE™

Selection of our books indexed in the Book Citation Index
in Web of Science™ Core Collection (BKCI)

Interested in publishing with us?
Contact book.department@intechopen.com

Numbers displayed above are based on latest data collected.
For more information visit www.intechopen.com

Meet the editor

Dr. Gemma Aiello was born in Aversa (CE), Italy, on October 24, 1964. In 1989, she graduated in Geological Sciences at the University of Naples "Federico II." In 1993, she earned a PhD degree in Sedimentary Geology at the University of Naples "Federico II," Department of Earth Sciences, Faculty of Geological Sciences. She completed a 2-year postdoctoral fellowship at the University of Naples "Federico II," and a CNR-CEE fellowship and several contracts at the Research Institute "Geomare Sud," CNR, Naples, Italy. Since 1998, she has been a full-time researcher at the Italian CNR. Dr. Aiello has 25 years of experience in the field of sedimentary geology, marine geology, and geophysics, participating in different research projects of the Italian National Research Council (CARG, Vector, Centri Regionali di Competenza). She was a contract professor of sedimentology and stratigraphy at the Parthenope University of Naples, Italy, and a teacher in formation courses for technicians in marine science and engineering in Naples, Italy.

Contents

Preface

This book contains four chapters dealing with the investigation of facies analysis and paleoecology, chemostratigraphy, and chronostratigraphy referring to paleoecological and facies analysis techniques and methodologies. The chapters pertain in particular to an Oligo-Miocene carbonate succession of the Persian Gulf (Asmari Formation), the chemostratigraphy of Paleozoic carbonates of Peninsular Malaysia through the integration of stratigraphic, sedimentologic, and geochemical data, and the chronostratigraphy of a small ice-dammed paleolake in Andorra (Spain), applying fast Fourier transform analysis, resulting in 6th-order stratigraphic cycles, which have outlined the occurrence of system tracts and unconformities controlled by glacio-eustasy. The chapters are separated into four main sections: (1) introduction; (2) facies analysis and paleoecology; (3) chemostratigraphy; and (4) chronostratigraphy. There is one chapter in the first section, introducing the stratigraphic setting of Paleozoic to Miocene deposits based on different stratigraphic methodologies, including facies analysis, paleoecology, chemostratigraphy, and chronostratigraphy. In the second section, there is one chapter dealing with the Oligocene-Miocene Asmari Formation, allowing for the recognition of several depositional environments based on sedimentological analysis, distribution of foraminifera, and micropaleontological study. In the third section, there is one chapter aimed at addressing research on the chemostratigraphy of cores, allowing for a significant increase in the stratigraphic knowledge existing on the Kinta Valley (Malaysia), coupled with extensive field work on Paleozoic carbonates. In the fourth section, there is a chapter dealing with the high-resolution chronostratigraphic setting of a paleolake located in Andorra (Spain) and the inference with the MIS2 isotopic stage of Atlantic and Mediterranean regions in the regional geological setting of the southeastern Pyrenees.

Introductory chapter

This chapter introduces the stratigraphic setting of Paleozoic to Miocene deposits based on different stratigraphic methodologies, including facies analysis, paleoecology, chemostratigraphy, and chronostratigraphy, applied in this book. Concepts and methods of facies analysis and paleoecology are discussed. Chemostratigraphy, a relatively young discipline in the field of stratigraphy, has been defined, coupled with recent attempts for its formalization in stratigraphy. The definitions of chronostratigraphy and geochronology, deeply revised in recent stratigraphic literature, are further clarified.

Facies analysis and paleoecology

The second chapter, "Paleoecology and Sedimentary Environments of the Oligo-Miocene Deposits of the Asmari Formation (Qeshm Island, SE Persian Gulf)" by Seyed Hadi Sajadi and Roya Fanati Rashidi, improves the geological knowledge of the Oligocene-Miocene Asmari Formation, allowing for the recognition of several depositional environments based on sedimentological analysis, distribution of foraminifera, and micropaleontological study.

Chemostratigraphy

The third chapter, "Chemostratigraphy of Paleozoic Carbonates in the Western Belt (Peninsular Malaysia): A Case Study on the Kinta Limestone" by Haylay Tsegab and Chow Weng Sum, takes the Kinta Valley (Malaysia) as an example, addressing stratigraphic research on the chemostratigraphy of cores, allowing for a significant increase in the stratigraphic knowledge existing in this area, coupled with extensive field work on Paleozoic carbonates.

Chronostratigraphy

The fourth chapter, "High Resolution Chronostratigraphy from an Ice-Dammed Palaeo-Lake in Andorra: MIS 2 Atlantic and Mediterranean Palaeo-Climate Inferences over the SE Pyrenees" by Turu Valenti, studies the high-resolution chronostratigraphic setting of a paleolake located in Andorra (Spain) and the inference with the MIS2 isotopic stage of Atlantic and Mediterranean regions in the regional geological setting of the southeastern Pyrenees.

I thank Mrs. Dolores Kuzelj, Author Service Manager of IntechOpen Science, Open Minds, who has contributed to this book on stratigraphy with competence and patience, following day after day the editorial activities and facilitating the publication of this book.

Dr. Gemma Aiello, PhD
Full-Time Researcher,
National Research Council of Italy (CNR),
Institute of Marine Sciences (ISMAR),
Section of Naples,
Naples, Italy

Section 1

Introduction

Introductory Chapter: An Introduction to the Stratigraphic Setting of Paleozoic to Miocene Deposits Based on Paleoecology, Facies Analysis, Chemostratigraphy, and Chronostratigraphy - Concepts and Meanings

Gemma Aiello

1. Introduction

This is the introductory chapter of the book "New insights into the stratigraphic setting of Paleozoic to Miocene deposits: case studies from the Persian Gulf, Peninsular Malaysia and south-eastern Pyrenees." In this chapter, the research themes studied in this book have been introduced referring to the paleoecological and facies analysis techniques and methodologies, pertaining, in particular, an Oligo-Miocene carbonate succession of the Persian Gulf (Asmari Formation), the chemostratigraphy of Paleozoic carbonates of Peninsular Malaysia through the integration of stratigraphic, sedimentologic, and geochemical data, and the chronostratigraphy of a small ice-dammed paleolake in Andorra, applying the FFT (Fast Fourier Transform) analysis, resulting in sixth-order stratigraphic cycles, which have outlined the occurrence of glacially controlled system tracts and unconformities.

The topic of the Asmari Formation and its depositional environments has been deeply studied [1–12]. Referring to its biostratigraphy, it was earlier outlined in the 1960s based on unpublished reports [11]. The application of isotopic stratigraphy has later proved that the sediments ascribed to the Miocene "Aquitanian" are, in fact, Late Oligocene, Chattian in age. This was proved by the application of Sr-isotope stratigraphy to cored sections from 10 Iranian oil fields and 14 outcrop sections, within the framework of a high resolution sequence stratigraphy study down to fourth order cycles. The Chattian/Aquitanian boundary is marked by a major faunal turnover, with the general extinction of *Archaias species* and *Miogypsinoides complanatus*. The age interpretation of the early, unpublished zonations has needed a deep revision and the establishment of an updated biozonation. The new zonation and the stratigraphic ranges of selected key species have been presented by Laursen et al., 2009 [11].

The isotopic stratigraphy based on strontium has constrained the stratigraphic setting of the Asmari Formation [8]. This formation, consisting of approximately 400 m of cyclic platform limestones and dolostones, with subordinate intervals of sandstones and shales, has been studied in the subsurface at several oil fields and in an outcrop section. The methods of Sr-isotope stratigraphy is suitable for dating these strata because of the fast rate of marine strontium ratio during the depositional processes (roughly 32–18 My). The profiles of age against depth in the four areas have shown a decrease from higher accumulation rates in the lower Asmari to lower rates in the middle-upper part of the formation. These changes reflect the dynamics of platform progradation, from early deposition along relatively high accommodation margin to slope settings and then, to conditions of lower accommodation on the shelf top [8]. The ages of the sequence boundaries have been estimated from the age-depth profiles at each locality, providing a framework for stratigraphic correlation. The depositional sequences have an average duration of 1–3 My, whereas the component cycles represent average time intervals of 100–300 ky.

On the other side, the Kinta limestones have been matter of previous studies, mainly referring to the depositional environments [13–16]. In the Kinta valley, they are composed of medium-to-dark gray, fine-grained, thinly bedded limestones, with preserved bedding planes and slump depositional features. The faunal content is quite scarce, except that some conodont faunas, while a high organic content is suggested from the dark color of the deposits. The sedimentological and facies analysis has suggested the occurrence of low energy, slope environment hosting the deposition of the Kinta limestones. The high organic content coupled with the lacking of benthic fauna has indicated a low-oxygen setting. On the other side, the Kinta limestones were dominated by mudstones interlayered with bedded cherts and perhaps were deposited in a slope environment with a significant contribution of pelagic deposits [13]. The geological evolution of the Kinta Valley has been recently outlined as characterized by both deposition and structural deformation [14]. During the Devonian, the deposition started, composed of alternating sandstones and mudstones, followed in the Carboniferous by fine-grained shales, which are, in turn, overlain by Permian limestones. During the Triassic-Early Jurassic, the intrusion of granites cut previously deposited carbonate deposits. The whole deposits are overlain by Quaternary alluvial deposits. An early compressional event and a late extensional event have been distinguished [14]. Folding and thrusting occurred during the compression, also controlling the granitic intrusion, which was fractured due to compressional deformation. The extensional tectonic event resulted from the individuation of normal faults, controlling the present-day drainage network evident from DEM analysis [14].

A high resolution biostratigraphy of the Kinta limestones has been later proposed based on conodonts sampled in three boreholes, composed of carbonate mudstones with shales and siltstones [15]. Nine diagnostic conodont genera and 28 age diagnostic conodont species have been identified. In particular, *Pseudopolygnathus triangulus triangulus* and *Declinognathodus noduliferus noduliferus* have indicated that the successions, pertaining to the Kinta limestones, range in age from the Upper Devonian to the Upper Carboniferous. Moreover, these data have provided clues to the Paleo-Tethys paleogeographic reconstruction and paleo-depositional conditions [15]. Recently, the deformational styles and the structural history of the Paleozoic limestones of the Kinta Valley have been defined by using remote sensing mapping, outcrop samples, and hand specimens [16]. An early extensional event has been identified, as marked from normal faults, while a compressional event was indicated by a set of strike-slip faults. The geologic evolution has been interpreted as an intra-basinal extension during Permo-Triassic times, which was followed by a Late Miocene to Quaternary tectonic uplift [16].

The high resolution chronostratigraphy of the paleo-lakes is a main research topic, which has been deeply studied by several authors [17–27]. In particular, the Ibate paleolake has shown a distal lacustrine environment with low-oxygen conditions in its bottom waters [17].

The occurrence of *Anacolosidites eosenonicus* sp. nov., combined with the lacking of *Steevesipollenites nativensis*, indicates a late Santonian age for the paleolake (ca. 84 Ma). This age is constrained by the occurrence of carbonized sclereids that are associated with the "Great Santonian Wildfire" recorded in coeval marine offshore strata of the Campos and Santos basins [17]. The palynological content, coupled with the occurrence of rhythmic deposits have indicated a Late Santonian age of these deposits. The age assignment is based on palynostratigraphic relationships established from a reliable biostratigraphic framework, based on integration of palynological and biostratigraphic data [17]. On the other side, the Qaidam lake represents an excellent example in order to study the interplay of climatic and tectonic controls on continental saline lakes [19]. Two main events of increase of salinity have been controlled by the climate during the Late Eocene since the Oligocene, while tectonic events have controlled the migration of the saline centers [19]. The accumulation of halites and their preservation were the result of a coupled control by active tectonism, in order to provide accommodation space and trigger a rapid subsidence.

The Navamuno peatbog system, located in western Spain, has been deeply studied [21]. During the Late Pleistocene, it was dammed by the Cuerpo de Hombre glacier and was fed by lateral meltwaters. This depression was then filled by glacio-lacustrine deposits. During the Holocene, its geologic evolution was controlled by a fluvial plain, controlling the episodes of shallow pond/peat bog sedimentation. An age model was constructed based on radiocarbon dating, allowing to interpret the environmental changes during the Late Glacial and the post-glacial [21]. Another representative paleolake is the Tangra Yumco, represented by a wide saline paleolake located on the Tibetan Plateau, which has been recently studied as a valuable example in order to reconstruct the climatic variations [23]. Micropaleontologic and sedimentologic data have been integrated with isotopic stratigraphy. Integrated stratigraphic information has allowed to reconstruct the geologic evolution of the paleolake during the last 17 ky [23]. The lake level was low at 17 ky BP, followed by a highstand phase at 8–9 ky BP. Since 2.5 ky, the paleolake remained stable regarding its level, with a short highstand-lowstand cycle around 2 ky [23]. These changes have been considered as good hints of paleo-climatic conditions in order to refine the paleo-climatic models in this area.

In this book, different case studies have been presented, respectively, located in the Persian Gulf, in the Peninsular Malaysia, and in the Andorra. To this aim, it should be useful to clarify their geological structure to put the studied cases in a proper geological setting [28–30]. The Persian Gulf is represented by an enclosed sea, limited from the western Arabian platform to the south and by the Zagros fold and thrust belt to the north-east. These mountains define the zone of convergence between the Arabian plate and the Eurasian plate and represent, perhaps, a tectonically active area. Since the last glacial maximum (18 ky BP), the sea level fluctuations in the Persian Gulf have been predicted in order to show their variability [28]. The paleo-shoreline reconstructions of the gulf have been compared with the general models of glacio-hydro-isostatic effects. Starting from the peak of the glaciations (14 ky), the Persian Gulf is free from the marine influence. The present shoreline of the Persian Gulf was reached about 6 ky ago, also controlling the evolution of the deltas of the rivers Euphrates, Tigris, and Kan [28]. In the Persian Gulf, the present-day water depths do not exceed 100 m, while the average water depths are of 35 m, suggesting that it was above the sea level during glacial times.

The geological setting of the Persian Gulf and the Oman Gulf has been studied by Ross et al. [29]. During Mesozoic times, the Arabian platform was formed by the Arabian Peninsula, by the Persian Gulf, by the south-western Iran, and by the eastern Iraq [29]. Significant geological processes outlined in this region include the deformation of the Musandam Peninsula during the Late Cretaceous and the

Middle Tertiary and the corresponding subduction processes, the collision of the Arabian platform and of the Eurasian plate, controlling the formation of the Zagros fold and thrust belt. This orogenesis has reduced the former platform to the Persian Gulf. This reduction was also controlled by the tectonic uplift of the Arabian Peninsula during the opening of the Red Sea and by saline tectonism [29]. During recent times, tectonics is still active in this complex region at the northern edge of the Gulf of Oman. Here, the Arabian plate has undergone subduction, while the Arabian and Eurasian plates lie in a collisional setting.

As a general rule, the Persian Gulf Basin represents a foreland basin, lying between the western Zagros fold and thrust belt, whose formation was controlled by the collision between the Arabian and the Eurasian plates [30]. An interesting topic is that the name "Persian Gulf" refers not only to the Persian Gulf but also to the Gulf of Oman, to the Straits of Hormuz, and to various outlets which are genetically related to the Arabian Sea. During the Early Triassic, the thermal subsidence and the stretching of the Arabian Plate started, resulting in extensional faulting and rifting of Zagros, opening the neo-Tethys sea. During the Late Cretaceous, a new tectonic phase controlled the beginning of the Alpine orogeny, resulting in major uplift and erosion, in addition to the closure of the Neo-Tethys sea [30]. During the Tertiary tectonic phase, the Late Alpine orogeny verified, resulting from the collision of the Arabian and Eurasian plates, resulting in the formation of the Zagros fold and thrust belt and then, the individuation of the foreland Persian Basin. Another main geodynamic event is represented by the opening of the Red Sea, about 25 My ago, resulting in the separation of the African and Arabian plates [30].

In this book, another important research topic is represented by the Peninsular Malaysia [31–36]. Three main tectonostratigraphic belts characterize these regions, respectively, the Western Peninsular Malaysia, the central Peninsular Malaysia, and the eastern Peninsular Malaysia. The oldest rocks can be found at the northwestern portion of the peninsula, while relatively younger rocks can be found toward the southeast. In the Peninsular Malaysia, the Upper Paleozoic and Mesozoic sequences have been studied in detail, regarding the structural and stratigraphic setting [32]. In particular, the Upper Paleozoic sequences have revealed several phases of folding coupled with the regional metamorphism, perhaps suggesting the occurrence of two main compressional events affecting the Peninsular Malaysia (Late Permian and Middle-Late Cretaceous) [32]. The Late Permian compressional event has controlled the intrusions of major plutons, cropping out in the eastern range. Harbury et al. [32] have suggested that the Permo-Triassic granites of the eastern belt have been separated from the granites cropping out in the main range due to crustal attenuation and subsidence during the Triassic and the Jurassic. I have found very clear on the geology of Peninsular Malaysia the study of Metcalfe [34]. This author has suggested that the aforementioned three belts occur based on different stratigraphic and structural settings, coupled with magmatism, geophysical signatures, and geologic evolution. The Western Belt is composed of the Sibumasu Terrane, derived from the margin of Gondwana during the Permian. The central and the eastern belts are composed of the Sukhothai Arc, formed during the Late Carboniferous-Early Permian on the Indochina continental margin [34]. During the Early Triassic, the collision between the Sibumasu and Sukhothai Arcs started, allowing for the formation of a foredeep basin and of an accretion complex. Granitic intrusions have cut the Western Belt and the Bentong-Raub suture zone. A back-arc basin (Sukhothai) opened during the Early Permian, collapsing and closing during the Middle-Late Triassic. In the Malay Peninsula, the marine deposition ended during the Late Triassic and red beds formed a cover sequence during the Cretaceous. A main tectonic and thermal event occurred during the Late Cretaceous, coupled with individuation of faults and granitic intrusion [34].

The third research topic of this book is represented by the geology of the Andorra (Spain), put in the regional context of the south-eastern Pyrenees [37–42]. The Andorra region is located in the central Pyrenees (Spain). This region has been strongly folded during the rotation of the Iberian Peninsula on the European plate. The stratigraphy of the Andorra region is characterized by the occurrence of rocks ranging in age from the Cambrian to the Ordovician, composed of conglomerates, limestones, phyllites, quarzites, and slates [40]. Moreover, gneiss and schist crop out in the cores of anticlines located in the north-eastern sector of the country. The occurrence of antiforms and anticlines is linked with shear zones including thrusts of metamorphosed sediments. In the south-eastern Andorra region, the Mt. S. Louis-Andorra Batholith crops out, controlling the metamorphism on its western edge.

A classical paper dealing with the Andorra's geology is that of Hartevelt [41]. The study region includes part of the Axial Zone, the Nogueras Zone, and the related marginal throughs. The outcropping formations, mapped with detail, range in age from the Cambro-Ordovician to the Pliocene. The detailed lithostratigraphy of this formation has allowed for the stratigraphic correlation with other regions of the Pyrenees. In this zone, the Hercynian orogenesis has controlled the formation of geological structures controlled by N-S trending stresses. A first tectonic phase has formed wide folds of kilometric extension, while the second one has controlled the formation of different compressional structures [41]. The thrust sequences in the eastern Pyrenees have been deeply investigated [42]. In this region, the Alpine thrusts involve both the basement and the sedimentary cover. Balanced cross sections have been constructed in order to restore the geometry of the thrusts and the propagation sequence, so resulting in a piggy-back sequence [42]. A duplex has been reconstructed, whose sole thrust is represented by the Vallfogona thrust, while the roof thrust owes its roots in the Axial Zone. Small antiforms have also been reconstructed, occurring as wide folds involving the higher sequences [42]. Casas et al. [43] have discussed the role of the Hercynian and Alpine thrusts in the Upper Paleozoic rocks of the Central and Eastern Pyrenees. The geological structure of the pre-Hercynian rocks of the Central and Eastern Pyrenees, forming the antiformal stack of the so-called Axial Zone, is characterized by coeval folds and thrusts, both Alpine and Hercynian. These thrusts separate sheets, ranging in age from the Upper Paleozoic to the Devonian, showing a different lithostratigraphy and geological structure [43]. Some examples have been shown in order to discuss the role of the Hercynian and Alpine thrusts in controlling the geological setting of the Pyrenees [43].

2. Facies analysis and paleoecology

In this book, the sedimentary environments and the paleoecology of the Oligo-Miocene deposits of the Asmari Formation have been reconstructed based on biostratigraphy, microfacies analysis, and facies analysis (see Chapter 1). Moreover, in Chapter 2, the facies analysis of Paleozoic carbonates drilled by three boreholes located in the Western Belt has been carried out. Perhaps, it should be useful to clarify some concepts and methods of facies analysis and paleoecology.

The stratigraphic analysis is mainly based on the field geological survey, on the measurement of stratigraphic sections and on the lithologic and paleontologic descriptions, with the aim to reconstruct the depositional environments and to correlate the stratigraphic sequences. A basic paper on facies analysis is that of Flugel [44], showing that every facies in a depositional setting is characterized by petrographic, geognostic, and paleontological characters, clearly different from the same characters of other facies occurring in the same geological period. The facies analysis needs interdisciplinary studies, as stated by Amanz Gressly in 1838 [45], showing that in the facies analysis, the sedimentologic, paleontologic, and geochemical data provide a basic information about the depositional environments, the lithogenesis, and the fossils.

In particular, the concept of facies needs to be recalled. It is a rocky body having distinct lithological, physical, and biological characteristics, allowing for its distinction from the adjacent rocky bodies. The concept of facies is usually referred to the whole characteristics of a sedimentary unit, including, the lithology, the grain-size, the sedimentary structures, the color, the composition, and the biogenic content [46]. A single facies does not indicate a single environment, but one or more geological processes through which the sediments have been deposited [47]. Perhaps, the environmental interpretation may be derived by the concept of facies association and by the integration of the physical characters of the deposits with the paleoecological ones. The facies associations are composed of several facies, occurring in combination and representing one or more depositional environments or facies groups, which are genetically related one to each other. Their shape is the cycle or the sequence, which is not a random vertical succession of facies. The facies associations are controlled by the Walther law, one of the most basic principles of stratigraphy. On the other side, a facies model is a general summary of depositional systems, including many single examples from recent sediments and old rocks.

Main criteria of facies analysis are briefly recalled [46, 47]. They include: (i) the mineralogic and petrographic composition, which gives information mainly on the provenance (relief, climate, and lithology of the source area), but also on the transport and on the diagenesis; (ii) the textural analysis, giving information on the provenance (shape), but mainly on the dynamics of transport and deposition; (iii) the fossil content, allowing for the dating and the correlation of deposits and giving paleoecological information and on the reworking; (iv) direction data, consisting of paleo-currents and paleo-slope models and dispersal of sediments, deduced from current lineations and depositional geometries; (v) geometry of the sedimentary bodies, derived through the synthesis of previous data and giving information on the depositional environments; and (vi) vertical sequential analysis, allowing for the determination of the relative depth fluctuations, the shoreline migrations, the growth and retreat of depositional systems, and the evolution of the sedimentary basins (basin analysis).

The paleoecology is represented by the study of the interactions between the organisms and the environments across the geological time scales and is linked with other disciplines, including the paleontology, the ecology, the climatology, and the biology [48]. It was born as a branch of the paleontology through the examination of the fossil and the ancient life environments. The main paleoecological approaches include: (i) the classic paleoecology, which is based on the fossils allowing for the reconstruction of the ancient ecosystems and uses the fossil remnants, such as the shells, the teeth, the pollens, and the seeds. A final result will be a paleo-environmental reconstruction. (ii) The evolutionary paleoecology, based on the holistic approach and using both the fossils and the physical and the chemical changes in the atmosphere, lithosphere, and hydrosphere in order to study the vulnerability and the resilience of species and environments. (iii) The community paleoecology, based on statistical methods and making use of physical models and computer analyses [48].

A main aim of the paleoecology is to construct a detailed model of the environments of life of the fossils, using the archives (represented by sedimentary sequences), the proxies (providing evidence of the biota and the related physical environments), and the chronology, allowing for the dating of events in the archive. Important proxies to carry out these reconstructions include the charcoal and pollens, particularly applied in paleolakes and peats. Some main paleoecological studies have been carried out in the Persian Gulf [49, 50]. Abdolmaleki and Tavakoli [49] have stated as the Permo-Triassic boundary represents one of the most important mass extinctions during the history of the earth, marking for a strong decrease of the living taxa. Important changes of depositional processes also occurred, forming anachronistic facies in whole earth. Anachronistic facies have been reported in the Early Triassic deposits of the Persian Gulf, consisting of microbial facies, composed of stromatolitic boundstones, oncoidal facies, and thrombolytic facies. The formation of these facies has also been controlled by the fluctuations in the $CaCO_3$ saturation level [49]. García-Ramos et al. [50] have evaluated the live-dead fidelity of the Mollusk assemblages in soft sediments of the carbonate tidal flats along the coasts of the Persian Gulf. Wide differences of this parameter have been controlled by the early cementation, lateral mixing, strong bioturbation, and low sedimentation rates. The obtained results have suggested that the average times in carbonate tidal flats are higher if compared with the times affecting the subtidal carbonate environments [50].

3. Chemostratigraphy

In this book, the chemostratigraphy of Paleozoic carbonates of the Western Belt (Peninsular Malaysia) has been studied (see Chapter 2). Perhaps, it should be useful to recall the chemostratigraphy as a branch of the integrated stratigraphy. Different stratigraphic methods are included in the integrated stratigraphy, including the chemostratigraphy, the isotopic stratigraphy, the oxygen isotopes, the carbon isotopes, the strontium isotopes, the orbital cyclostratigraphy, the response of the climate system to the orbital forcing, the orbital forcing and the sedimentary environments, the identification of cyclical features, and the spectral analysis of time series. Particular attention must be given to the methods of absolute dating and to the geological time scale. The chemostratigraphy (chemical stratigraphy) is based on the study of the chemical variations in the sedimentary successions with the aim to reconstruct the stratigraphic relationships [51–55]. It is based on the principle that the chemical signatures may be used as fossil groups or lithological groups in order to establish the stratigraphic relationships between the rocky layers. The types of chemical variations may be summarized [51]. Colorimetric variations among the strata may be detected in some stratigraphic sequences, triggered by the content of metals of transition incorporated during the deposition. Other colorimetric variations may be controlled by variations in the content of organic carbon in the deposits. The development of new techniques of analysis, including the electronic microprobe and the X-ray fluorescence, has facilitated the chemical analysis of the deposits, coupled with the geochemistry of the stable isotopes. In particular, the variability of the oxygen in the carbonate shells of foraminifera represents a proxy for the temperatures of the ocean during the geological past [56–57]. Recently, there were some attempts to formalize the chemostratigraphy as a standard method of stratigraphy [54–55], but this discipline is too young and many efforts need to be made again.

4. Chronostratigraphy

In this book, a chronostratigraphic reconstruction of lacustrine deposits located in Spain has been carried out (see Chapter 3). For this reason, it is useful to recall some chronostratigraphic concepts. Recently, the definitions of chronostratigraphy and geochronology have been deeply revised [58]. The realignment of the two terms has been proposed, contemporaneously solving the problem if the Geological Time Scale must have single or double time units. This discussion must be carried out based on the use of the Geological Time Scale (GTS) reported in the International

Stratigraphic Chart of the International Commission of Stratigraphy and its units. It must be taken into account that the last version of the International Stratigraphic Chart has been published in 2018 [59]. The most used units are represented by the geological periods of the geochronology (Triassic, Jurassic, for instance) and the chronostratigraphic systems on which they are based. These systems are composed of series and stages, while the periods, epochs, and stages are referred to time intervals during which the deposition of strata occurred. Therefore, there is a double hierarchy of chronostratigraphic units (time/rocks), which have been used to indicate rocky strata contemporaneously deposited and time intervals (geochronologic) used to indicate intervals during which geological processes occurred, including the evolution, the extinction, the deformation, and the transgression/regression, for instance [58]. In the meaning of this paper, the geochronology indicates the timing and the age of main geological events of the earth's history (such as a glaciations or a mass extinction). Moreover, it refers to the methods of numerical dating.

On the other side, the definition of chronostratigraphy is quite different. It includes the whole range of the stratigraphic disciplines, such as the magnetostratigraphy, the chemostratigraphy, the sequence stratigraphy, the cyclostratigraphy, and the radiometric dating [60–64]. The main aims of the chronostratigraphy include both the establishment of the time relations of regional successions and the definition of a GSSP (Global Boundary Stratotype Section and Point). In the realignment proposed by Zalasiewicz et al. [58], the chronostratigraphic (time/rock) and geochronologic (time) units have been, respectively, defined as it follows: (i) eonothem (Phanerozoic, for instance); (ii) erathem (Mesozoic, for instance); (iii) system (Cretaceous, for instance); (iv) series (Upper Cretaceous, for instance); (v) stage (Cenomanian, for instance); (vi) eon (Phanerozoic, for instance); (vii) era (Mesozoic, for instance); (viii) period (Cretaceous, for instance); (ix) epoch (Late Cretaceous, for instance); and (x) age (Cenomanian, for instance). The proposed method is to use the chronostratigraphic units in reference to layered rocks and to use the geochronologic units in reference to time and phenomena associated to the rocks [58].

5. Outline of this book

This book examines different stratigraphic studies regarding the fields of facies analysis and paleoecology, chemostratigraphy, and chronostratigraphy focusing on several applications, including the paleoecology and the sedimentary environments of the Asmari Formation (south-eastern Persian Gulf), allowing for the recognition of two assemblage zones based on foraminifera, indicating an age ranging between the Chattian and the Aquitanian, while facies analysis has indicated a depositional environment of carbonate ramp, the chemostratigraphy of Paleozoic carbonates in the Western Belt through the integration of stratigraphic, sedimentological, and geochemical data on three boreholes, indicating that the variations of major elements is directly related to the lithofacies types in the study samples and finally the application of the chronostratigraphic chart through a Fast Fourier Transform (FFT) analysis on lacustrine deposits of Andorra (Spain), indicating the occurrence of 6th order stratigraphic cycles, genetically related to high frequency sea level fluctuations during the Late Pleistocene.

This book contains four chapters, as follows:

Chapter 1 [Introductory Chapter: An Introduction to the stratigraphic setting of Paleozoic to Miocene deposits based on paleoecology, facies analysis, chemostratigraphy, and chronostratigraphy: Concepts and Meanings].

Chapter 2 [Paleoecology and Sedimentary Environment of the Oligocene-Miocene (Asmari Formation) deposits, in Qeshm Island, SE Persian Gulf].

Chapter 3 [Chemostratigraphy of Paleozoic carbonates in the Western Belt, Peninsular Malaysia; case study from the Kinta Valley].

Chapter 4 [High-resolution chronostratigraphy from an ice-dammed paleolake in Andorra: MIS 2 Atlantic and Mediterranean paleoclimate inferences over the SE Pyrenees].

Conflict of interest

I declare that there is no conflict of interest.

Author details

Gemma Aiello
Istituto di Scienze Marine (ISMAR), Consiglio Nazionale delle Ricerche (CNR), Napoli, Italy

*Address all correspondence to: gemma.aiello@iamc.cnr.it

IntechOpen

References

[1] Vaziri-Moghaddam H, Kimiagari M, Taheri A. Depositional environment and sequence stratigraphy of the Oligo-Miocene Asmari formation in SW Iran. Facies. 2006;**52**:41. DOI: 10.1007/s10347-005-0018-0

[2] Hamedani A, Torabi H, Piller W, Mandic O, Steininger FF, Wielandt U, et al. Oligo-Miocene Sections from Zagros Foreland Basins of Central Iran. Abstract, 18th IAS Reg Meet Sediment1997. pp. 155-156

[3] Schuster F, Wielandt U. Oligocene and early Miocene coral faunas from Iran: Palaeoecology and palaeogeography. International Journal of Earth Sciences. 1999;**88**:571-581

[4] Seyrafian A. Microfacies and depositional environments of the Asmari formation at Dehdez area (a correlation across central Zagros Basin). Carbonates and Evaporites. 2000;**15**:22-48

[5] Seyrafian A, Hamedani A. Microfacies and depositional environment of the upper Asmari formation (Burdigalian), north-central Zagros Basin, Iran. Neues Jahrbuch für Geologie und Paläontologie-Abhandlungen. 1998;**210**:129-141

[6] Seyrafian A, Hamedani A. Microfacies and palaeoenvironmental interpretation of the lower Asmari formation, Oligocene, north central Zagros Basin, Iran. Neues Jahrbuch fur Geologie und Palaontologie-Monatshefte. 2003;**3**:164-167

[7] Seyrafian A, Vaziri H, Torabi H. Biostratigraphy of the Asmari formation, Burujen area, Iran. Journal of Sciences, Islamic Republic of Iran. 1996;**7**:31-47

[8] Ehrenberg SN, Pickard AH, Laursen GV, Monibi S, Mossadegh ZH, Svana TA, et al. Strontium isotope stratigraphy of the Asmari formation (Oligocene-lower Miocene), SW Iran. Journal of Petroleum Geology. 2007;**30**(2):107-128

[9] Aqrawi AM, Keramati M, Ehrenberg SN, Pickard N, Moallemi A, Svna T, et al. Origin of dolomite in the Asmari formation (Oligocene-lower Miocene), Dezful embayment, SW Iran. Journal of Petroleum Geology. 2006;**29**(4):381-402

[10] van Buchem FSP, Allan TL, Laursen GV, Loftpour A, Moallemi S, Monibi H, et al. Regional stratigraphic architecture and reservoir types of the Oligo-Miocene deposits in the Dezful embayment (Asmari and Pabdeh formations) SW Iran. Geological Society of London, Special Publication. 2010;**329**:219-263

[11] Laursen GV, Monibi S, Pickard A, Hoissenev A, Vincent B, Hamon Y, et al. The Asmari Formation Revisited: Changed Stratigraphic Allocation and New Biozonation. Shiraz: Eage; 2009

[12] Seyrafian A. Carbonates and Evaporites. 2000;**15**:121. DOI: 10.1007/BF03175819

[13] Gebretsadik HT, Hunter AW, Sum CW. Depositional Environment of the Kinta Limestone, Western Peninsular Malaysia. AAPG Datapages/Search and Discovery Article #90194©. Istanbul, Turkey: International Conference & Exhibition; 2014

[14] Meng CC, Sautter B, Pubellier M, Menier D, Sum CW, Kadir A. A geological features of the Kinta Valley. Platform. 2014;**10**(2):2-14

[15] Gebretsadik HT, Sum CW, Gatovsky A, Hunter AA, Talib J, Kassa S. Higher-resolution biostratigraphy for the Kinta limestone and an implication for continuous sedimentation in the

Paleo-Tethys, Western Belt of peninsular Malaysia. Turkish Journal of Earth Sciences. 2017;**26**:377-394

[16] Meng CC, Pubellier M, Abdeldayem A, Sum CW. Deformation styles and structural history of the Paleozoic limestone, Kinta Valley, Perak, Malaysia. Bulletin of the Geological Society of Malaysia. 2016;**62**:37-45

[17] Mitsuru A, Brito D. The Ibaté paleolake in SE Brazil: Record of an exceptional late Santonian palynoflora with multiple significance (chronostratigraphy, paleoecology and paleophytogeography). Cretaceous Research. 2018;**84**:264-285

[18] Tourloukis V, Muttoni G, Karkanas P, Monesi E, Scardia G, Panago E, et al. Magnetostratigraphic and chronostratigraphic constraints on the Marathousa lower Paleolithic site and the middle Pleistocene deposits of the megalopolis basin, Greece. Quaternary International. 2018;**497**:47-64

[19] Guo P, Liu C, Yu M, Ma D, Wang P, Wang K, et al. Paleosalinity evolution of the Paleogene perennial Qaidam lake on the Tibetan plateau: Climatic vs. tectonic control. International Journal of Earth Sciences. 2018;**107**(5):1641-1656

[20] Borner A, Hrynowiecka A, Stachowicz R, Niska RM, Del Hoyo M, Kuztetsov V, et al. Paleoecological investigations and ^{230}U/Th dating of the Eemian interglacial peat sequence from Neubrandenburg-Hinterste Muhle (Mecklenburg-Western Pomerania). Quaternary International. 2018;**467**(A):62-78

[21] Turu V, Carrasco R, Pedraza J, Ros X, Zapata B, Soriano-Lopez J, et al. Late glacial and post-glacial deposits of the Navamuno peatbog (Iberian central system): Chronology and palaeoenvironmental implications. Quaternary International. 2018;**470**(A):82-95

[22] Lowry DP, Morrill C. Is the last glacial maximum a reverse analog for future hydroclimate changes in the Americas? Climate Dynamics. August 2018:1-21. DOI: 10.1007/s00382-018-4385-y

[23] Alivernini A, Akita LG, Ahlborn M, Borner N, Haberzetti T, Kasper T, et al. Ostracod-based reconstruction of late quaternary lake level changes within the Tangra Yumco lake system (southern Tibetan plateau). Journal of Quaternary Sciences. 2018;**33**:713-720

[24] Barrett S, Drescher-Schneider R, Stanberger R, Spotl C. Evaluation of the regional vegetation and climate in the eastern Alps (Austria) during MIS 3-4 based on pollen analysis of the classical Baumkirchen paleolake sequence. Quaternary Research. 2018;**90**(1):153-163

[25] Nicoll K. A revised chronology for Pleistocene paleolakes and middle stone age-middle Paleolithic cultural activity at Bir Tirfawi-Bir Sahara in the Egyptian Sahara. Quaternary International. 2018;**463**(A):18-28

[26] Hrynowieckca A, Zarski M, Jakubowski G, et al. Eemian and Vistulian (Weichselian) palaeoenvironmental changes: A multi-proxy study of sediments and mammal remains from the Lawy palaeolake (Eastern Poland). Quaternary International. 2018;**467**(A):131-146

[27] Cartier R, Brisset E, Guiter F, Sylvestre F, Tachikawa K, Anthony E, et al. Multiproxy analyses of Lake Allos reveal synchronicity and divergence in geosystem dynamics during the late glacial/Holocene in the Alps. Quaternary Science Reviews;**186**:60-77

[28] Lambeck K. Shoreline reconstructions for the Persian Gulf since the last glacial maximum. Earth and Planetary Science Letters. 1996;**142**:43-57

[29] Ross DA, Uchupi E, White R. The geology of the Persian Gulf of Oman region: A synthesis. Reviews of Geophysics and Space Physics. 1986;**24**(537)

[30] Konyuhov AI, Maleki B. The Persian Gulf Basin: Geological history, sedimentary formations and petroleum potential. Lithology and Mineral Resources. 2006;**41**(4):344-361

[31] Cocks LRM, Fortey RA, Lee CP. A review of lower and middle Paleozoic biostratigraphy in west peninsular Malaysia and southern Thailand in its context within the Sibumasu Terrane. Journal of Asian Earth Sciences. 2005;**24**:703-717

[32] Harbury NA, Jones ME, Audley-Charles MG, Metcalfe I, Mohamed KR. Structural evolution of the peninsular Malaysia. Journal of the Geological Society. 1990;**147**:11-26

[33] Lee CP, Mohamed SL, Kamaludin H, Bahari MN, Rashidah K. Stratigraphic Lexicon of Malaysia. Malaysia: Geological Society of Malaysia; 2004

[34] Metcalfe I. Tectonic evolution of the Malay peninsula. Journal of Asian Earth Sciences. 2013;**76**:195-213

[35] Hazad F, Azman A, Ghani A, Hua Lo C. Arc related dioritic-granodioritic magmatism from southeastern peninsular Malaysia and its tectonic implication. Cretaceous Research. 2019;**95**:208-224

[36] Baioumy H, Ulfa Y, Nawawi M, Padmanabahn E, Anuar N. Mineralogy and geochemistry of Paleozoic black shales from peninsular Malaysia: Implications for their origin and maturation. International Journal of Coal Geology;**165**:90-105

[37] Turu V, Calvet M, Bordonau J, Gunnell Y, Delmas M, Vilaplana JM, et al. Did Pyrenean glaciers dance to the beat of global climatic events? Evidence from the Wurmian sequence stratigraphy of an ice-dammed palaeolake depocentre in Andorra. In: Hughes PD, Woodward JC, editors. Quaternary Glaciation in the Mediterranean Mountains. Vol. 433(1). Geological Society of London, Special P Publication; 2017. pp. 111-136

[38] Sancho C, Arenas C, Pardo G, Pena-Monné JL, Rhodes EJ, Bartolomé M, et al. Glacio-lacustrine deposits formed in an ice-dammed tributary valley in the south-Central Pyrenees: New evidence for late Pleistocene climate. Sedimentary Geology. 2018;**366**:47-66

[39] Jalut G, Turu V, Dedoubat JJ, Otto T, Ezquerra J, Fontugne M, et al. Palaeoenvironmental studies in NW Iberia (Cantabrian range): Vegetation history and synthetic approach of the last deglaciation phases in the western Mediterranean. Palaeogeography, Palaeoclimatology, Palaeoecology. 2010;**297**:330-350

[40] Available from: https://en.wikipedia.org/wiki/Geology_of_Andorra

[41] Hartevelt JJ. Geology of the upper Segre and Valira valleys, Central Pyrenees, Andorra, Spain. Leidse Geologische Mededelingen. 1970;**45**:167-236

[42] Munoz JA, Martinez A, Verges J. Thrust sequences in the eastern Spanish Pyrenees. Journal of Structural Geology. 1986;**8**(3-4):399-405

[43] Casas J, Domingo F, Poblet J, Soler A. On the role of the Hercynian and alpine thrusts in the upper Paleozoic rocks of the central and eastern Pyrenees. Geodinamica Acta. 1989;**3**(2):135-147

[44] Flugel E. Introduction to facies analysis. In: Microfacies Analysis of Limestones. Berlin Heidelberg: Springer; 1982. pp. 1-26

[45] Gressly A. Observations géologiques sur le Jura Soleurois. Vol. 2. Neuchâtel: N. Denkschr. Allgem. Schweiz. Ges. Naturwiss; 1838

[46] Ricci Lucchi F. Sedimentologia. Bologna, Italia: Cooperativa Libraria Universitaria Editrice Bologna; 1978

[47] Mutti E, Ricci Lucchi F. Le torbiditi dell'Appennino settentrionale: Introduzione all'analisi di facies. Vol. 11. Memorie della Società Geologica Italiana; 1972. pp. 161-199

[48] Available from: https://en.wikipedia.org/wiki/Paleoecology

[49] Abdolmaleki J, Tavakoli V. Anachronistic facies in the early Triassic successions of the Persian Gulf and its palaeoenvironmental reconstruction. Palaeogeography, Palaeoclimatology, Palaeoecology. 2016;**446**:213-224

[50] García-Ramos DA, Albano PG, Harzhauser M, Piller WE, Zuschin M. High dead-live mismatch in richness of molluscan assemblages from carbonate tidal flats in the Persian (Arabian) gulf. Palaeogeography, Palaeoclimatology, Palaeoecology. 2016;**457**:98-108

[51] Berger WH, Vincent E. Chemostratigraphy and biostratigraphic correlation: Exercises in systematic stratigraphy. Oceanologica Acta. 1981:115-127

[52] Renard M, Corbin JC, Daux V, Emmanuel L, Baudin F, Tamburini F. Chapter 3: Chemostratigraphy. In: Rey J, Galeotti S, editors. Stratigraphy: Terminology and Practice. Paris, France; Editions Ophrys. pp. 41-52. ISBN 978-2-7108-0910-4

[53] Prothero DR, Schwab F. Section IV: Stratigraphy. Chapter 17: Geophysical and chemostratigraphic correlation. In: Freeman WH, editor.

macmillan international, London, UK: Sedimentary Geology. 3rd ed. 2014. ISBN 978-1-4292-3155-8

[54] Ramkumar M, editor. Chemostratigraphy: Concepts, Techniques and Applications. Elsevier; Amsterdam, The Netherlands. 2015. 530 p. ISBN 978-0-12-419968-2

[55] Ramkumar M. Toward standardization of terminologies and recognition of chemostratigraphy as a formal stratigraphic method. In: Ramkumar M, editor. Chemostratigraphy: Concepts, Techniques and Applications. Elsevier; Amsterdam, The Netherlands. 2015. pp. 1-21. ISBN 978-0-12-419968-2

[56] Mortyn PG, Martinez Botì MA. Planktonic foraminifera and their proxies for the reconstruction of surface-ocean climate parameters. Contributions to Science. 2007;**3**:371-383

[57] Pearson PN. Oxygen isotopes in foraminifera: Overview and historical review. In: Ivany L, Huber B, editors. Reconstructing Earth's Deep Time Climate–The State of the Art in 2012. Paleontological Society Short Course; 3 November, 2012. The Paleontological Special Papers. Vol. 18. 2012. pp. 1-38

[58] Zalasiewicz J, Cita MB, Hilgen F, Pratt BR, Strasser A, Thierry J, et al. Chronostratigraphy and geochronology: A proposed realignment. GSA Today. 2013;**23**(3):4-8. DOI: 10.1130/GSATG160.A.1

[59] Available from: http://www.stratigraphy.org/index.php/ics-chart-timescale

[60] Strasser A, Hilgen F, Heckel PH. Cyclostratigraphy—Concepts, definitions and applications. Newsletters on Stratigraphy. 2006;**42**:75-114

[61] Weissert H, Joachimiski M, Sarntheim M. Chemostratigraphy. Newsletters on Stratigraphy. 2008;**42**:145-179

[62] Langereis CG, Krijgsman W, Muttoni G, Menning M. Magnetostratigraphy—Concepts, definitions and applications. Newsletters on Stratigraphy. 2010;**43**:207-233

[63] Catuneanu O, Galloway WE, Kendall C, Miall AD, Posamentier HW, Strasser A, et al. Sequence stratigraphy: Methodology and nomenclature. Newsletters on Stratigraphy. 2011;**44**:173-245

[64] Gradstein FM, Ogg JG, Schmitz MD, Ogg GM. The Geologic Time Scale 2012. Oxford: Elsevier; 2012. p. 1144

Section 2

Facies Analysis and Paleoecology

Chapter 2

Paleoecology and Sedimentary Environments of the Oligo-Miocene Deposits of the Asmari Formation (Qeshm Island, SE Persian Gulf)

Seyed Hadi Sajadi and Roya Fanati Rashidi

Abstract

The Asmari Formation is composed of limestones, marly limestones, and marls, whose subsurface thickness in this region is about 148 m. Two assemblage zones have been recognized through the distribution of large foraminifera in the study area, indicating a Late Oligocene (Chattian)-Early Miocene (Aquitanian) age. The gradual facies changes and the lacking of turbiditic deposits show that the Asmari Formation was deposited in a carbonate ramp environment. Based on the depositional textures and petrographical studies, characterizing gradual shallowing upward trends of an open marine carbonate ramp, three distinct depositional settings have been recognized: lagoon, barrier, and open marine. MF1 was characterized by the occurrence of hyaline benthic and planktonic foraminifera representing distal middle ramp and below the storm wave base of other ramp. Paleolatitudinal reconstructions based on skeletal grains suggest that carbonate sedimentation of the Asmari Formation took place in tropical waters within the photic zone.

Keywords: Asmari Formation, microfacies, paleoecology, benthic foraminifera, Oligocene-Miocene, Qeshm Island

1. Introduction

This chapter deals with the Asmari Formation (one of the best known carbonate reservoirs in the world) [1], an Oligocene-Miocene carbonates succession cropping out in the south-eastern Zagros basin, southern Iran (**Figure 1**). At the type section outcropping in Tang-e Gel-e Tursh (Valley of Sour Earth), which is located on the south-western flank of the Kuh-e Asmari anticline, the Asmari Formation mainly consists of limestones, dolomitic limestones, and argillaceous limestones [3, 4], having an average thickness of 314 m. In the Qeshm Island, the Asmari shallow marine limestone is located in the subsurface and was deposited over the Pabdeh Formation with a gradational stratigraphic contact. The contact with the overlying Gachsaran Formation (i.e., evaporitic rocks) is conformable and gradual (**Figure 2**). This formation is present in the most part of the Zagros basin, and its lithology is characterized by limestones, dolomitic limestones, dolomites, and

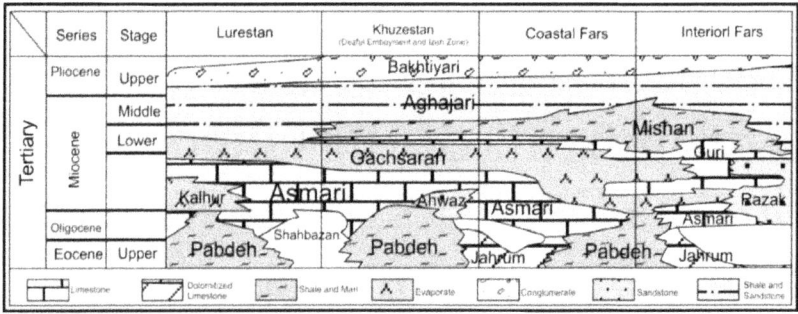

Figure 1.
Cenozoic stratigraphic correlation chart of the Iranian sector of the Zagros Basin, after James and Wynd [2].

marly limestones. Some anhydrite (Kalhur Member) and lithic and limy sandstones (Ahwaz Member) also occur within the Asmari Formation [3, 4]. Previous studies have focused on biostratigraphy and lithostratigraphy of the Asmari Formation and were originally defined in primary works [5–8]. Later, other researchers have introduced the microfaunal characteristics and the assemblage zones for the Asmari Formation [2, 9, 10]. More recent studies of the Asmari Formation have been conducted on facies and sedimentary environment [8, 11–17]. Referring to the biostratigraphy of the Asmari Formation, it was earlier outlined in the 1960s based on unpublished reports [18]. The application of the isotopic stratigraphy has later proved that the sediments ascribed to the Miocene "Aquitanian" are in fact Late Oligocene, Chattian in age. This was proved by the application of Sr-isotope stratigraphy to cored sections from 10 Iranian oil fields and 14 outcrop sections, within the framework of a high-resolution sequence stratigraphic study down to fourth order cycles. The Chattian/Aquitanian boundary is marked by a major faunal turnover, with the general extinction of *Archaias* species and *Miogypsinoides complanatus*. Main insights on the stratigraphic setting of the Asmari Formation have been given from the strontium isotopic stratigraphy [19]. The Asmari Formation

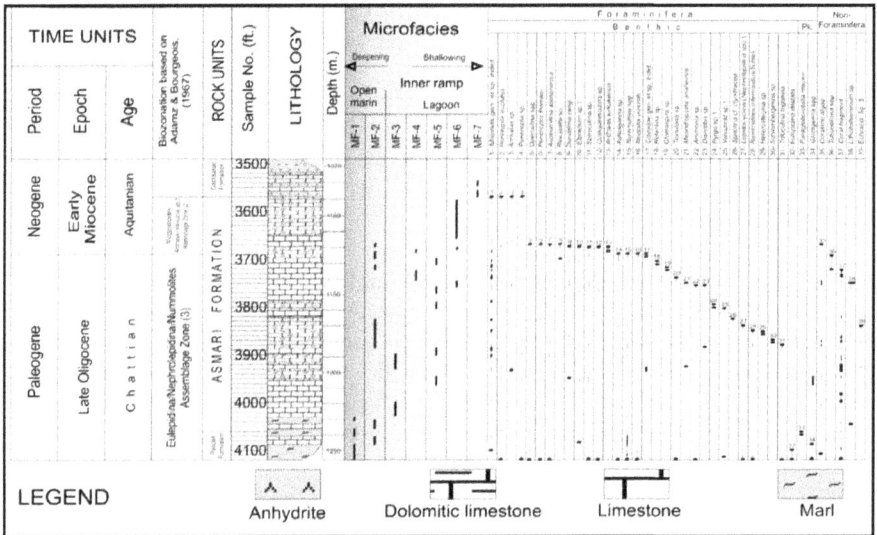

Figure 2.
Lithostratigraphy column, microfacies, benthic and planktonic foraminifers' distribution and biozonation of the Asmari Formation at Qeshm Island (well no. 2).

has been studied in the subsurface at the Bibi Hakimeh, Marun, and Ahwaz oilfields and in an outcrop section from the Khaviz anticline. It consists of approximately 400 m of cyclic platform limestones and dolostones with subordinate intervals of sandstone and shale. The method of Sr-isotope stratigraphy is well suited for dating these strata because of the rapid rate of change of marine strontium ratio during Asmari deposition (roughly 32–18 Ma) and the common presence of well-preserved macrofossils. Profiles of age against depth in the four areas show a decrease from higher stratigraphic accumulation rates in the lower Asmari to lower rates in the middle to upper part of the formation. There is also a trend toward less open marine depositional conditions and increasing early dolomitization and anhydrite abundance above the lower part of the formation. These changes reflect the dynamics of platform progradation across the areas studied, from early deposition along relatively high accommodation margin to slope settings to later conditions of lower accommodation on the shelf top. Ages of sequence boundaries have been estimated from the age-depth profiles at each locality, providing a framework for stratigraphic correlation. The Asmari deposition began in early Rupelian time (34–33 Ma) in the Bibi Hakimeh area, when basinal marly facies accumulated in the north-western sector of the study areas. The depositional sequences have durations of 13 Ma, whereas the component cycles represent average time intervals of 100–300 Ky. This chapter reports on the subsurface sedimentological study of the Asmari Formation, whose results have been correlated and compared for a better geologic comprehension of the outcrops of the Asmari Formation in the adjacent areas. The objectives of this study are (1) a description of the facies and their distribution on the Oligocene-Miocene carbonate platform and (2) an interpretation of the paleoenvironmental features based on the assemblages of benthic hyaline and imperforate foraminifera.

2. Geological setting

The Zagros Basin is the second largest basin in the Middle East and is defined by a 7–14-km thick succession of coverage sediments deposited over a region located along the north-northeast edge of the Arabian plate. This basin was part of the stable Gondwana supercontinent in the Paleozoic era and of a passive margin in the Mesozoic era, and it became a site of plate convergence and formation of thrust belts in the Cenozoic era [20]. The Zagros Fold-and-Thrust Belt of Iran is a result of the Alpine orogenic events [21, 22] in the Alpine-Himalayan mountain range. It extends in a NW-SE direction from eastern Turkey to the strait of Hormoz in southern Iran. The tectonic activity of this area was entirely due to the convergence of the Arabian and Eurasian continents. After the closure of the Neo-Tethys basin, during late Oligocene-early Miocene times, the Zagros basin was gradually narrowed and the Asmari Formation was deposited with a lithology including lithic sandstone (Ahwaz Member) and evaporites (Kalhur Member) [1, 23]. The maximum thickness of the Asmari Formation is found in the north-eastern corner of the Dezful Embayment. On the basis of the lateral facies variations, the Iranian Zagros fold-thrust belt is divided into different tectono-stratigraphic domains, which are from SE to NW: the Fars Province or eastern Zagros, the Khuzestan province or Central Zagros, and finally the Lurestan Province or Western Zagros [3, 4] (**Figure 3b**). Also, from south-west to north-east of the Zagros basin, there are the Zagros folded belt, folded and thrusted belt, and High Zagros and crush zone [25–28]. The Hormozgan Province is located in southern Iran and is part of Zagros Folded belt. This region is accompanied by NW-SE, W-E, and N-S trending simple anticlines and synclines with very great thickness of Fars Group deposits (Gachsaran, Mishan, Aghajari, and Bakhtiari Formations) and presence of 118 salt plugs. So, for these specific features, Motiei [3, 4] called this area as the "Bandar Abbas Hinterland" (**Figure 3**).

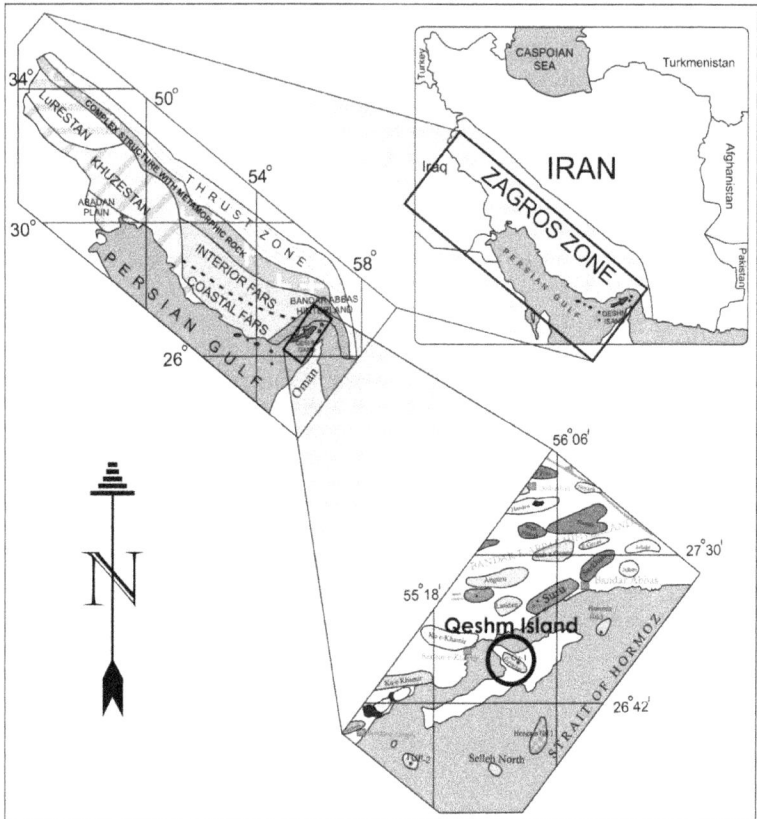

Figure 3.
Geological location and geological map of the studied section, modified after geological map [24].

3. Methods and study area

This study involves one stratigraphic subsurface section from the Asmari Formation. The study area is located at Qeshm Island, southern Iran (**Figure 3c**). The lithologies and the microfacies types were classified and described according to Dunham [29]. Some samples from the underlying Pabdeh and overlying Gachsaran Formations were also analyzed for boundaries distinction. A total of 60 thin sections of the cores and cuttings have been analyzed under the microscope for biostratigraphy and facies. Petrographic studies were carried out for facies analysis and paleoenvironmental reconstruction of the Asmari Formation. Facies have been determined for each paleoenvironment according to carbonate grain types, textures, and interpretation of functional morphology of small and larger foraminifers. Biostratigraphy has been determined based on the well-known benthic foraminifera biozones of Adams and Bourgeois [30].

4. Result

4.1 Biostratigraphy

Biostratigraphic criteria of the Asmari Formation were established by Wynd [10] and reviewed by Adams and Bourgeois [30] in unpublished reports only. Biozonation and age determinations in the study area are based on benthic foraminifera biozonation of Adams and Bourgeois [30]. From the base to the top, two foraminiferal assemblages have been recognized and were discussed as it follows:

Assemblage I. This assemblage corresponds to the *Eulepidina-Nephrolepidina-Nummulites* Assemblage Zone (3) [30]. The assemblage is considered to be Chattian in age. The most diagnostic species include Miliolids gen. et sp. Indet., *Peneroplis evolutus*, *Archaias* sp., *Peneroplis* sp., *Operculina* spp., *Peneroplis thomasi*, *Austrollina asmariensis*, *Reussella* sp., *Dendritina rangi*, *Elphidium* sp. 1, *Spiroculina* sp., *Quinqueloculina* sp., *Asterigerina* sp., *Nummulites* spp., *Neorotalia viennoti*, Cibicidae gen. et sp. Indet, *Archaias kirkukensis*, *Hetererilina* sp., *Glomospira* sp., *Textularia* sp., *Meandropsina anahensis*, *Ammonia* sp., *Discorbis* sp., *Pyrgo* sp. 1, *Valvulinid* sp. 1, *Spirolina* cf. *clyndracea*, *Lepidocyclina* (*Nephrolepidina* spp.), *Nummulites intermedius/fichteli*, *Heterostegina* sp., *Schlombergerina* sp., *Triloculina trigonula*, *Eulepidina dilatata*, *Rotalia* sp., *Bolivina* sp., *Paragloborotalia mayeri*, and *Globigerina* spp.

Assemblage II. This assemblage corresponds to the *Miogypsinoides-Archaias-Valvulinid* sp. 1 Assemblage Zone (2) [30]. The assemblage is considered to be Aquitanian in age. The most important foraminifera in this assemblage are Miliolids gen. et sp. Indet., *Peneroplis evolutus*, *Archaias* sp., *Peneroplis* sp., *Operculina* spp., *Peneroplis thomasi*, *Austrollina asmariensis*, *Reussella* sp., *Dendritina rangi*, *Elphidium* sp. 1, *Spiroculina* sp., *Quinqueloculina* sp., and *Archaias kirkukensis*.

4.2 Microfacies analysis

The microfacies analysis of the Asmari Formation in the study area has resulted in the definition of seven types of facies, which characterize the platform development. Each microfacies exhibits typical skeletal and non-skeletal components and related sedimentary textures. These facies are related to the three depositional settings (lagoon, barrier, and open marine) of inner, middle, and outer portions of a carbonate platform (**Figure 4**). Since the Asmari Formation overlies the Pabdeh Formation and conformably underlies the Gachsaran Formation, some samples from the Pabdeh and Gachsaran Formations have also been studied. The general environmental interpretation of the microfacies is discussed in the following paragraphs.

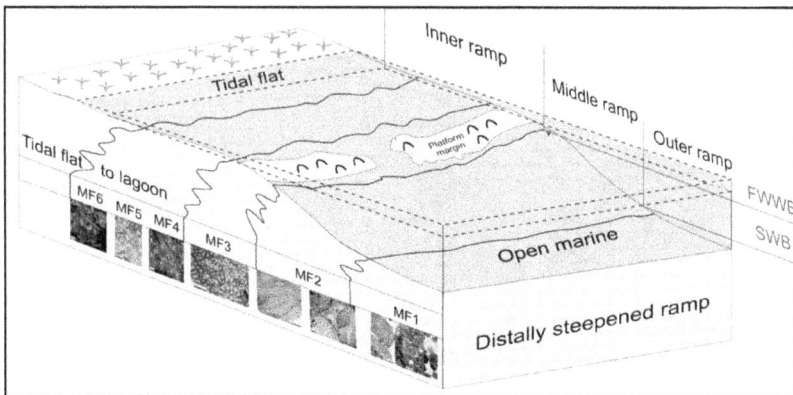

Figure 4.
Depositional model for the carbonate platform of the Asmari Formation at the southeast of Zagros basin, Qeshm Island [31].

4.2.1 MF1 marl facies

There are intercalations of marl across the section, but this facies mainly occurs in the lower part of the succession (**Figure 5A–D**). They are gray to green marl and contain benthic (miliolids, *Nummulites*, *Neorotalia*, *Elphidium*, *Operculina*, *Amphistegina* and textularids) and planktonic (*Paragloborotalia mayeri* and *Globigerina* spp.) foraminifera. The planktonic foraminifera occur at the base of the succession, where the boundary between the Pabdeh and Asmari Formations is located [32].

4.2.1.1 Interpretation

The features of benthic faunas and the stratigraphic relationships with the other microfacies suggest that the marly facies was deposited in an open lagoon

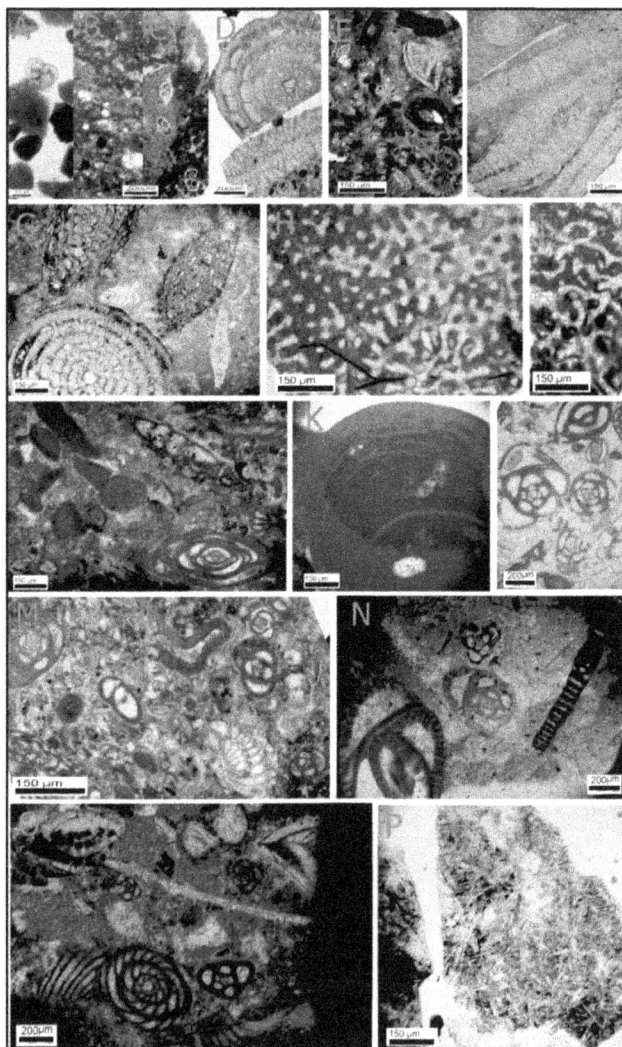

Figure 5.
Microfacies types of the Asmari Formation. (A–D) MF1, Marl facies. (E–G) MF2, bioclastic Lepidocyclinidae, Nummulitidae, Neorotalia, Wackestone-packstone. (H and I) MF3, coral boundstone. (J and K) MF4, miliolids corallinacea bioclastic wackestone. (L and M) MF5, miliolids bioclastic wackestone. (N and O) MF6, imperforate foraminifera bioclast wackestone-packstone. (P) MF7, evaporite.

with a normal-salinity water, but the coexistence of planktonic and some benthic (Nummulitidae) foraminifera in the base of the Asmari marls and marly limestones has suggested that this facies was deposited in calm, low-energy hydrodynamic, and deep normal-salinity water, which indicates a deposition below the storm wave base [33–36].

4.2.2 MF2 bioclastic wackestone-packstone with Lepidocyclinidae, Nummulitidae, and Neorotalia

This microfacies is composed of grain-supported texture with abundant large benthic foraminifera (**Figure 5E–G**). The foraminiferal assemblage is represented by numerous large benthic perforate foraminifera such as Lepidocyclinidae and Nummulitidae (*Nummulites* and *Operculina*). Other components such as *Astigerina* and red algae are rare. Due to changes in the type of fauna in some samples, the name of this facies changes to bioclastic wackestone-packstone with Lepidocyclinidae, Nummulitidae and Neorotalia. The biostratigraphic distribution and paleoenvironmental model of the Asmari Formation in this stratigraphic interval are most prominent in the lower parts of the Asmari Formation [37].

4.2.2.1 Interpretation

It consists of gray marly limestone beds. The combination of micritic matrix and abundance of typical open marine fauna including large Nummulitidae, Lepidocyclinidae and Neorotalia suggest a low-medium energy, open marine environment. Other bioclasts such as red algae and shell fragments are rare. This microfacies shows an environment between the storm wave base and fair-weather wave base (FWWB) [35, 36]. The presence of large *Nummulites* and lepidocyclinids suggests that this microfacies took place in relatively deep water and was formed in the lower photic/oligophotic zone in a distal middle ramp [22, 38–50].

4.2.3 MF3 coral boundstone

This facies is characterized by the abundance of scleractinian and massive coral colonies (**Figure 5H and I**).

4.2.3.1 Interpretation

This microfacies is interpreted to be formed by in situ organisms as an organic reef (Bioherm) in margin of platform and was located above the fair-weather wave base (FWWB) [36].

4.2.4 MF4 miliolids corallinacea bioclastic wackestone

Miliolids, coralline red algae and coral are dominating components in this microfacies (**Figure 5J and K**). Other bioclasts are rare but include *Peneroplis* and dendritic fragments. The textures are wackestones.

4.2.4.1 Interpretation

The MF5 represent low- to medium-energy open lagoon shallow subtidal environments, but there is different from MF4 by their texture and grain composition.

Depositional textures, fauna and stratigraphic position took place in warm, euphotic and shallow water, with low to moderate energy conditions, in a semi-restricted lagoon. This area is located within inner carbonate platform setting [32].

The presence of well-preserved coralline algae indicates a relatively quiet-water environment with a stable substrate and low sedimentation rates [51]. The associations of miliolids within this facies support the additional interpretation of a relatively protected environment, probably the inner part of a platform [52].

4.2.5 MF5 miliolids bioclastic wackestone

This facies is characterized by the dominant presence of small benthic foraminifera (miliolids) (**Figure 5L** and **M**). Other components such as *Peneroplis*, *Elphidium*, bryozoan and extraclasts are rare. The matrix is fine-grained micrite.

4.2.5.1 Interpretation

This facies is characterized by low diversity skeletal fauna and was deposited in a restricted low-energy lagoonal environment. There is a low biotic diversity of fauna, which shows a high-stressed habitat in very shallow restricted areas, where great fluctuations in salinity and temperature probably occurred [52].

4.2.6 MF6 imperforate foraminifera bioclast wackestone-packstone

The main elements of this microfacies are skeletal and non-skeletal components (**Figure 5N** and **O**). The skeletal components include a high diversity of imperforate foraminifera in grain-supported textures and several genera of benthic foraminifera (*Austrotrillina*, *Archaias*, *Peneroplis*, *Meandropsina*, *Elphidium*, *Dendritina* and miliolids). Peloids are rare, and other minor biota consists of particles of bryozoans and corals.

4.2.6.1 Interpretation

The occurrence of large number of porcelain imperforate foraminiferal tests may point to the depositional environment being slightly hypersaline [15]. These deposits include different textures ranging from wackestone to packstone. Some porcelain imperforate foraminifera (*Peneroplis* and *Archaias*) live in recent tropical and subtropical shallow water environments [53]. Textural characteristics and prolific porcelain foraminifera suggest that a medium-to-high energy portion of a restricted lagoon with a nearby tidal flat sedimentary environment prevailed [17]. Such an assemblage can be associated with an inner ramp environment [1, 17, 35, 36, 53, 54].

4.2.7 MF7 evaporite

Anhydrite and gypsum facies have been observed in the upper part of the Asmari Formation, which represents the beginning of the Gachsaran Formation (**Figure 5P**). The first anhydrite has been deposited above the marly limestones with a sharp contact.

4.2.7.1 Interpretation

Considering the deposition of anhydrite implies that the depositional environment became isolated from the open marine at that time, which has allowed for the concentration and submarine precipitation of salt. The thickness of the evaporate deposits indicates that they are submarine deposits formed in an isolated saline basin. A eustatic sea level fall is one of the most likely causes. This event took place around the early Miocene (Aquitanian), and its stratigraphic expression was recorded at the boundary of the Asmari and Gachsaran Formations. Based on Ehrenberg et al. [19], this anhydrite is exposed at the top of the Asmari Formation and indicates the Oligocene-Miocene boundary. Ehrenberg et al. [19] noted that strontium dates got from anhydrite formed as an evaporate rather than as a later diagenetic product.

5. Discussion

5.1 Sedimentary development of the Oligocene-Miocene Fars sub-basin

Planktonic and benthic foraminifera and non-foraminifera distribution of the Oligocene deposits can represent the type of sedimentary environment, adopted from joint project of French and Iran Oil Company [55] (**Figure 6**). During the Paleogene, Pabdeh (basinal marls and argillaceous limestones) Formation was deposited in the middle and on both sides of the Zagros basinal axis [3] (**Figure 1**). The shallow marine limestones of the Asmari Formation were deposited above the Pabdeh Formation in the section of this study (**Figure 1**). During the Rupelian and early Chattian, outer ramp facies (Pabdeh

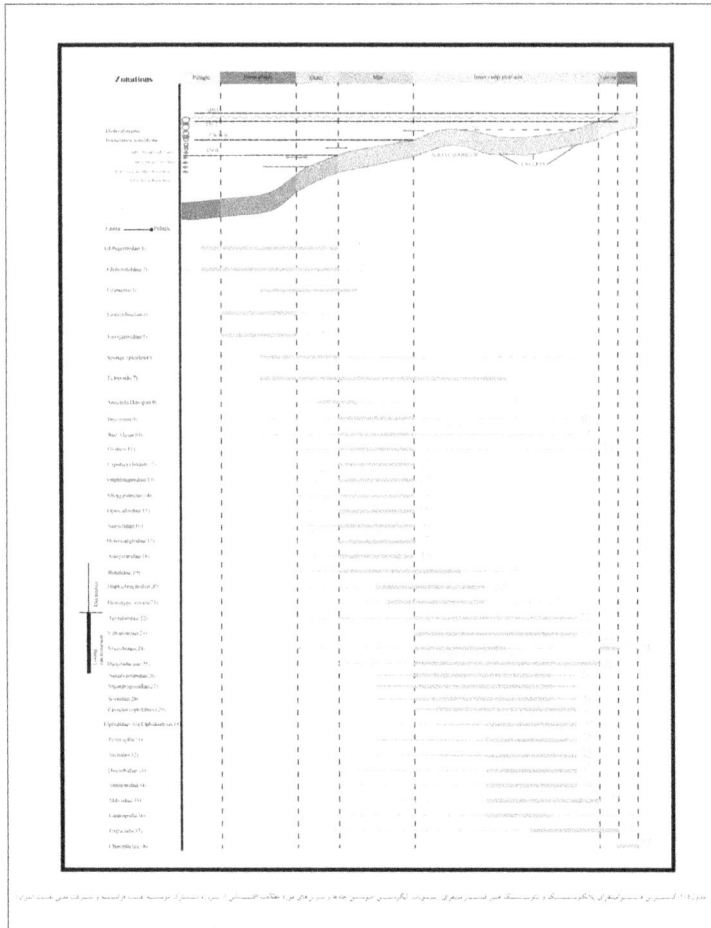

Figure 6.
Foraminifera and non-foraminifera distribution of the Oligocene deposits, adopted from joint project of French and Iran Oil Company [55].

Formation) was predominant at the Qeshm section (well no. 3) (**Figure 2**). This is visible in the lower part of the Asmari Formation. So, the Chattian sediments of the Asmari Formation in this section gradationally overlie the Pabdeh Formation. Indeed, Chattian basin in this time restricted by shallow subtidal environments.

5.2 Paleoecology

Large benthic foraminifera (such as Nummulitidae) produced great amount of carbonates during the Early and Middle Paleogene. In the Oligocene, euphotic conditions prevailed and carbonate production related to these foraminifers (especially *Nummulites*) declined [56]. Larger perforate forms are represented by *Amphistegina*, nummulitids and lepidocyclinids. Perforate foraminifera that live in shallow waters are characterized by hyaline walls and so protect themselves from ultraviolet light by producing very thick, lamellate test walls to prevent photo inhibition of symbiotic algae within the test in bright sunlight. These large forms are the most important indicators for constructing paleoenvironmental models in the warm, shallow marine environments [42]. The presence of these large and flat forms (Lepidocyclinidae and Nummulitidae) in the lower part of Asmari Formation, in comparison with analogues in the modern platform, allowed interpreting these sediments as having been deposited in the lower photic zone [41–45, 48]. In contrast, coralline red algae communities become dominant, as most phototrophic carbonate producers thrive in shallow marine environments [56], especially through Early Miocene to Tortonian [57]. Coralline red algae and large benthic foraminifera (*Nummulites, Operculina, Lepidocyclina, Archaias, Peneroplis* and *Dendritina*) are the most significant and dominant biota in the Asmari Formation at the study area. Other components such as corals, bryozoan and echinoderms are present within the matrix. The distribution of larger foraminifera and coralline red algae are largely dependent on the salinity, depth, light, temperature and climate, nutrients, effect of hydrodynamic energy and flow substrate on the biostrate and dispersion of taxa [13, 58]. Small benthonic foraminifera are common locally and include porcellaneous (miliolids) and perforated (rotaliids) forms. Rotaliids are dominated by *Neorotalia viennoti* specimens. Larger foraminifera represented by the porcellaneous imperforate tests such as *Archaias* and *Peneroplis* may point to the depositional environment being within the photic zone in tropical carbonate platforms and slightly hypersaline [17, 35, 37, 54]. Flatter tests and thinner test walls with increasing water depth reflect decreased light levels at greater depths or perhaps poor water transparency in shallow waters [40]. These test shapes reflect adaptation to low hydrodynamic energy. Some biogenic components such as miliolids indicate stress conditions within restricted environments. Miliolids-dominated benthic foraminiferal assemblages reflect a decreased circulation and probably a reduced oxygen contents or euryhaline conditions. Miliolids are found in a variety of very shallow, hyposaline to hypersaline environments or are even common in the sand shoal environments of normal salinities [59, 60] and are generally taken as evidence of restricted lagoon [53].

5.3 Depositional environments

Three depositional environments have been identified in the Oligocene-Miocene succession of the Qeshm Island, on the basis of the biostratigraphic content and of the facies relationships (**Figure 6**). These include lagoon, barrier and open marine (**Figure 4**). These three environments are represented by seven microfacies types (MF1: distal middle ramp and below the storm wave base of other ramp,

MF2: deeper fair water wave base of a middle ramp setting and MF 3–6: shallow water setting of an inner ramp influenced by wave and tide processes). Carbonate ramp environments are characterized by (1) the inner ramp, between the upper shore face and fair weather wave base, (2) the middle ramp, between fair weather wave base and storm wave base and (3) the outer ramp, below normal storm wave base down to the basin plain [61]. Inner ramp deposits represent marginal marine deposits indicative of open lagoon and protected lagoon. In the restricted lagoon environment, faunal diversity is low and normal marine faunae are lacking, except for imperforate benthic foraminifera such as miliolids and *Dendritina*, which indicate quite conditions. A large number of porcellaneous imperforates points to somewhat hypersaline waters [33, 52]. The presence of imperforate foraminifera that include *Archaias*, *Peneroplis*, *Dendritina*, *Meandropsina*, *Austrotrillina* and miliolids indicates a low-energy, upper photic, shallow lagoonal depositional environment. The large porcellaneous foraminifera types such as *Archaias*, *Peneroplis* and *Dendritina* are present in MF 6. The occurrence of *Archaias* and *Peneroplis* is typical of recent tropical and subtropical shallow water environments [46, 62] and are characteristics of the upper part of the upper photic zone (inner ramp). Furthermore, these large porcellaneous foraminifera are also common fossils in the Mesozoic and Cenozoic neritic sediments [57]. And also, inner ramp deposits represent a wider spectrum of marginal marine deposits, indicative of a high-energy reef (MF 3). The middle ramp setting is represented by the medium to fine grained foraminiferal bioclastic wackestones-packstones, dominated by assemblages of larger foraminifera with perforate walls such as *Amphistegina*, *Operculina*, and *Nummulites* (**Figure 5**). The faunal association suggests that the depositional environment was situated in the mesophotic to oligophotic zone [48, 63]. Open lagoon shallow subtidal environments are characterized by microfacies types that include mixed open marine bioclasts (such as red algae, echinoids and corals) and protected environment bioclasts (such as miliolids). The diversity association of skeletal components represents a shallow subtidal environment, with optimal conditions as regards salinity and water circulation. The change in larger foraminiferal fauna from porcellaneous imperforated to hyaline perforated forms points to a decrease in water transparency [38]. The microfacies 1 and 2 are subject to an open marine environment of a proximal outer ramp and middle ramp, respectively. More common components of the microfacies 1 is biota association, such as large benthic foraminifera (Lepidocyclinidae, *Nummulites* and *Operculina*), small benthic foraminifera (*Neorotalia*), coralline red algae, which is dominated in lower photic zone. Moreover, the red algae association with these larger foraminifera places the middle ramp in an oligophotic to mesophotic zone [48, 53, 57, 63, 64].

6. Conclusions

The Oligo-Miocene Asmari Formation is a thick sequence of shallow water carbonates and is widespread in the Zagros basin. The subsurface section of the Asmari Formation in the south-eastern part of the Zagros and Qeshm Islands has allowed to recognize different depositional environments based on the sedimentological analysis, on the distribution of the foraminifera and on the microfacies studies. The occurrence of large foraminifera (*Nummulites*, *Operculina*, *Lepidocyclina*, *Archaias*, and *Peneroplis*), coralline red algae, coral debris and fragments of Echinoderms, Mollusks and Bryozoans has evidenced that a high nutrient stability in an oligothrophic to mesothrophic condition existed during the deposition of the Asmari Formation. Based on the occurrence of these fossils, two assemblage zones (*Eulepidina-Nephrolepidina-Nummulites* Assemblage Zone and

Miogypsinoides-Archaias-Valvulinid sp. 1 Assemblage Zone) have been recognized, and the Asmari carbonate in the study area is Chattian-Aquitanian in age. Based on the occurrence of skeletal (large benthic foraminifera and coralline red algae) and non-skeletal components, the following environmental and palaeoecological implications have been defined for the Asmari depositional environment at the Qeshm Island, southern Bandar Abbas Hinterland. Based on components and texture, seven microfacies types have been recognised and grouped into three depositional environments, corresponding to inner, middle and outer carbonate ramp. The microfacies 1 and 2 were deposited in an open marine environment of a proximal outer ramp and middle ramp, respectively. The microfacies 3–6 belong to an inner ramp/platform environment. These assemblages of the Asmari Formation suggest that the carbonate sedimentation took place in tropical waters in oligotrophic to slightly mesotrophic conditions.

Acknowledgements

The studies were supported by National Iranian Oil Company (NIOC). The authors wish to thank the Exploration Directorate (NIOC) for financial support and permission to publish this research.

Author details

Seyed Hadi Sajadi[1*] and Roya Fanati Rashidi[2*]

1 Department of Geology, North Tehran Branch, Islamic Azad University, Tehran, Iran

2 Department of Geology, Science and Research Branch, Islamic Azad University, Tehran, Iran

*Address all correspondence to: h.sajadi10@gmail.com and roya_fanati@yahoo.com

IntechOpen

References

[1] Vaziri-Moghaddam H, Seyrafian A, Taheri A, Motiei H. Oligocene-Miocene ramp system (Asmari Formation) in the NW of the Zagros basin, Iran: Microfacies, paleoenvironment and depositional sequence. Revista Mexicana de Ciencias Geológicas. 2010;**27**:56-71

[2] James GA, Wynd JG. Stratigraphic nomenclature of Iranian Oil Consurtium agreement area. AAPG Bulletin. 1965;**49**:2182-2245

[3] Motiei H. Stratigraphy of Zagros. Treatise on the Geology of Iran. Ministry of Mines and Metals. Tehran: Geological Survey of Iran; 1993. p. 1

[4] Motiei H. Zagros Stratigraphy. In: Plan for Codifying Book. State Geology Organization, National Iranian Oil Company (NIOC); 1993

[5] Busk HG, Mayo HT. Some notes on the geology of the Persian oilfields. Journal Institute Petroleum Technology. 1918;**5**(17):5-26

[6] Richardson RK. The geology and oil measures of Southwest Persia. Journal of the Institute Petroleum Technology. 1924;**10**(43):256-296

[7] Van Boeckh HDE, Lees GM, Richardson FDS. Contribution to the stratigraphy and tectonics of the Iranian ranges. In: Gregory JW, editor. The Structure of Asia. London: Methuen and Co.; 1929. pp. 58-177

[8] Thomas NA. The Asmari Limestone of southwest Iran. National Iranian Oil Company; 1948. Report 706, unpublished

[9] Jalali MR.Stratigraphy of Zagros Basin. National Iranian Oil Company, Exploration and Production Division; 1987

[10] Wynd JG. Biofacies of the Iranian Oil Consortium Agreement area. In: Unpublished Report. Iranian Oil Operating Companies, Geological and Exploration Division; 1965

[11] Rahmani Z, Vaziri-Moghaddam H, Taheri A, Ghabeishavi A. A model for the paleoenvironmental distribution of larger foraminifera of Oligocene–Miocene carbonate rocks at Khaviz Anticline, Zagros Basin, SW Iran. Historical Biology. 2009;**21**(3-4):215-227

[12] Sadeghi R, Vaziri-Moghaddam H, Taheri A. Microfacies and sedimentary environment of the Oligocene sequence (Asmari Formation) in Fars sub-basin, Zagros Mountains, Southwest Iran. Facies. 2011;**57**(3):431-446

[13] Sooltanian N, Seyrafian A, vaziri-Moghaddam H. Biostratigraphy and paleo-ecological implications in microfacies of the Asmari Formation (Oligocene), Naura anticline (interior Fars of the Zagros Basin), Iran. Carbonates and Evaporites. 2011;**26**:167-180

[14] Taheri A, Vaziri-Moghaddam H, Seyrafian A. Relationships between foraminiferal assemblages and depositional sequences in Jahrum Formation, Ardal area (Zagros Basin, SW Iran). Historical Biology. 2008;**20**(3):191-201

[15] Vaziri-Moghaddam H, Kalanat B, Taheri A. Sequence stratigraphy and depositional environment of the Oligocene deposits at Firozabad section, southwest of Iran based on microfacies analysis. JGeope (Geopersia). 2011;**1**(1):71-82

[16] Vaziri-Moghaddam H, Kimiagari M, Taheri A. Depositional environment and sequence stratigraphy of the Oligo-Miocene Asmari Formation in SW Iran. Facies. 2005:1-11

[17] Vaziri-Moghaddam H, Kimiagari M, Taheri A. Depositional environment and sequence stratigraphy of the Oligocene—Miocene Asmari formation in SW Iran. Facies. 2006;**52**:4151

[18] Laursen GV, Monibi S, Allan TL, Pickard NAH, Hosseiney A, Vincent B, et al. The Asmari Formation revisited: Changed stratigraphic allocation and new biozonation. In: First International Petroleum Conference and Exhibition, Shiraz, Iran. 2009

[19] Ehrenberg SN, Pickard NAH, Laursen GV, Monibi S, Mossadegh ZK, Svånå TA, et al. Strontium isotope stratigraphy of the Asmari Formation (Oligocene–Lower Miocene), SW Iran. Journal of Petroleum Geology. 2007;**30**:107-128

[20] Bahroudi A, Koyi Hemin A. Tectono-sedimentary framework of the Gachsaran Formation in the Zagros foreland basin. Marine and Petroleum Geology. 2004;**21**:1295-1310

[21] Ricou LE, Braud J, Brunn JH. Le Zagros Mem Hors. Ser Soc Geol. 1977;**8**:33-52

[22] Sadeghi A, Vaziri-Moghaddam H, Taheri A. Biostratigraphy and paleoecology of the OligoMiocene succession in Fars and Khuzestan areas (Zagros Basin, SW Iran). Historical Biology. 2009;**21**(1-2):17-31

[23] Ahmadhadi F, Lacombe O, Marc Daniel J. Early reactivation of basement faults in central Zagros (SW Iran), evidence from pre-folding fracture populations in Asmari Formation and lower tertiary paleogeography. In: Lacombe O, Lave J, Roure F, Verges J, editors. Thrust Belts and Foreland Basins. Berlin: Springer; 2007. pp. 205-228

[24] Geological Division. Zagros Structures. National Iranian Oil Company (NIOC), Exploration Directorate; 2004

[25] Falcon NL. Southern Iran: Zagros mountains. In: Spencer A, editors. Mesozoic–Cenozoic Orogenic Belts. Geological Society, London, Special Publication. 1974;**41**:99-211

[26] Sepehr M, Cosgrove JW. Structural framework of the Zagros fold thrust belt, Iran. Marine and Petroleum Geology. 2004;**21**:829-843

[27] Sherkati S, Letouzey J. Variation of structural style and basin evolution in the central Zagros (Izeh zone and Dezful Embayment), Iran. Marine and Petroleum Geology. 2004;**21**:535-554

[28] Fakhari MD, Axen GJ, Horton BK, Hassanzadeh J, Amini A. Revised age of proximal deposits in the Zagros foreland basin and implications for Cenozoic evolution of the high Zagros. Techtonophysics. 2008;**451**:170-185

[29] Dunham RJ. Classification of carbonate rocks according to their depositional texture. In: Ham WE, editor. Classification of Carbonate Rocks. A Symposium: AAPG Bulletin. 1962. pp. 108-121

[30] Adams CG, Bourgeois E. Asmari biostratigraphy. In: Unpublished Report. Geological and Exploration Division, Iranian Oil Offshore Company; 1967

[31] Read JF. Carbonate margins of passive (extensional) continental margins: Types, characteristics and evolution. Tectonophysics. 1982;**81**:195-212

[32] Rahmani Z, Vaziri-Moghaddam H, Taheri A. Facies diestribution and paleoecology of the Guri member of the Mishan Formation, in Lar area, Fars province, SW Iran. Iranian Journal of Science & Technology, Transaction A, Printed in the Islamic Republic of Iran. 2010;**34**(A3):257-266

[33] Buxton MWN, Pedley HM. A standardized model for Thethyan

tertiary carbonates ramps. Journal of Geological Society London. 1989;**146**:746-748

[34] Cosovic V, Drobne K, Moro A. Paleoenvironmental model for Eocene foraminiferal limestones of the Adriatic carbonate platform (Istrian Peninsula). Facies. 2004;**50**:61-75

[35] Flugel E. Microfacies Analysis of Limestones. Analysis, Interpretation and Application. Berlin: Springer; 2004. p. 976

[36] Wilson JL. Carbonate Facies in Geologic History. Berlin, Heidelberg, New York: Springer; 1975. p. 471

[37] Amirshahkarami M, Vaziri-Moghaddam H, Taheri A. Paleoenvironmental model and sequence stratigraphy of the Asmari Formation in Southwest Iran. Historical Biology. 2007;**19**(2):173-183

[38] Barattolo F, Bassi D, Romero R. Upper Eocene larger foraminiferal-coralline algal facies from the Klokova Mountain (south continental Greece). Facies. 2007;**53**:361-375

[39] Bassi D, Hottinger L, Nebelsick H. Larger foraminifera from the Upper Oligocene of the Venetian area, Northeast Italy. Palaeontology. 2007;**50**(4):845-868

[40] Beavington-Penney SJ, Racey A. Ecology of extant nummulitids and other larger benthic foraminifera: Applications in paleoenvironmental analysis. Earth-Science Reviews. 2004;**67**(2):19-265

[41] Geel T. Recognition of stratigraphic sequence in carbonate platform and slope: Empirical models based on microfacies analysis of Paleogene deposits in southeastern Spain. Palaeogeography Palaeoclimatology Palaeoecology. 2000;**155**:211-238

[42] Hallock P. Symbiont bearing foraminifera. In: Sen Gupta BK, editor. Modern Foraminifera. Dordrecht: Kluwer; 1999. pp. 123-139

[43] Hohenegger J. Remarks on the distribution of larger foraminifera (Protozoa) from Palau (western Carolines). In: Aoyama T, editor. The Progress Report of the 1995 Survey of the Research Project, Man and the Environment in Micronesia. Kagoshima University Research Center for the Pacific Islands, Occasional Papers. Vol. 32. 1996. pp. 19-45

[44] Hottinger L. Répartition comparée des grands foraminifères de la mer Rouge et de l'Océan Indien. Annali dell'Università di Ferrara; 1980;(6):35-51

[45] Hottinger L. Processes determining the distribution of larger foraminifera in space and time. Utrecht Micropaleontological Bulletins. 1983;**30**:239-253

[46] Leutenegger S. Symbiosis in benthic foraminifera, specificity and host adaptations. Journal of Foraminiferal Research. 1984;**14**:16-35

[47] Nebelsick JH, Rasswer M, Bassi D. Facies dynamic in Eocene to Oligocene Circumalpine carbonates. Facies. 2005;**51**:197-216

[48] Pomar L. Ecological control of sedimentary accommodation: Evolution from a carbonate ramp to rimmed shelf, Upper Miocene, Balearic Islands. Palaeogeography Palaeoclimatology Palaeoecology. 2001a;**175**:249-272

[49] Reiss Z, Hottinger L. The Gulf of Aqaba. In: Ecological Micropaleontology. Vol. 354. Berlin: Springer; 1984

[50] Romero J, Caus E, Rosell J. A model for the palaeoenvironmental distribution of larger foraminifera

based on late middle Eocene deposits on the margin of the south Pyrenean basin (NE Spain). Palaeogeography Palaeoclimatology Palaeoecology. 2002;**179**:43-56

[51] Nebelsick JH, Bassi D. Diversity, growth forms and taphonomy, key factors controlling the fabric of coralline algae dominated shelf carbonates. In: Insalaco E, Skelton PW, Palmer TJ, editors. Carbonate platform system: Components and interaction. Geological Society, London, Special Publications. 2000;**178**:89107

[52] Fournier F, Montaggioni L, Borgomano J. Paleoenvironments and high-frequency cyclicity from Cenozoic southeast Asian shallow-water carbonates, a case study from the Oligo-Miocene buildups of Malampaya (offshore Palawan, Philippines). Marine and Petroleum Geology. 2004;**21**:1-21

[53] Brandano M, Frezza V, Tomassetti L, Cuffaro M. Heterozoan carbonates in oligotrophic tropical waters: The Attard member of the lower coralline limestone formation (Upper Oligocene, Malta). Palaeogeography Palaeoclimatology Palaeoecology. 2008;**272**:110

[54] Flugel E. Microfacies Analysis of Limestone. Vol. 633. Berlin: Springer; 1982

[55] IFP-INOP. Sequence Stratigraphy of Oligocene in Dezful Area, Southwest Iran. In: Joint Project between French and Iranian Oil Company. Unpublished Report. 2006

[56] Pedley M, Carannante G. Cool-water carbonates ramps: A review. In: Pedley M, Caranante G, editors. Cool-water carbonates: Depositional systems and paleoenvironmental controls. Geological Society, London, Special Publications. 2006;**255**:1-9

[57] Brandano M, Frezza V, Tomassetti L, Cuffaro M. Heterozoan carbonates in oligotrophic tropical water, the Attard member of the lower coralline limestone formation (Upper Oligocene, Malta). Palaeogeography, Palaeoclimatology, Palaeoecology. 2009;**56**:1138-1158

[58] Murray JWL. Ecology and Paleoecology of Benthic Foraminifera. Vol. 397. Harlow: Longman; 1991

[59] Brasier MD. Ecology of recent sediment-dwelling and phytal foraminifera from the lagoons of Barbuda, West Indies. Journal of Foraminiferal Research. 1975a;**5**:42-46

[60] Brasier MD. The ecology and distribution of recent foraminifera from the reefs and shoals around Barbuda, West Indies. Journal of Foraminiferal Research. 1975b;**5**:193-210

[61] Burchette TP, Wright VP. Carbonate ramp depositional systems. Sedimentary Geology. 1992;**79**:3-57

[62] Lee JJ. Fine structure of the rhodophycean Porphyridium purpureum in situ in *Peneroplis pertusus* (Forskal) and *P. acicularis* (Batsch) and in axenic culture. Journal of Foraminiferal Research. 1990;**20**:162-169

[63] Hottinger L. Shallow benthic foraminiferal assemblages as signals for depth of their deposition and their limitations. Bulletin of the Geological Society of France. Société Géologique de France. 1997;**168**(4):491-505

[64] Brandano M, Corda L. Nutrients, sea level and tectonics: Constrains for the facies architecture of a Miocene carbonate ramp in Central Italy. Terra Nova. 2002;**14**:257-262

Section 3

Chemostratigraphy

Chapter 3

Chemostratigraphy of Paleozoic Carbonates in the Western Belt (Peninsular Malaysia): A Case Study on the Kinta Limestone

Haylay Tsegab and Chow Weng Sum

Abstract

The Peninsular Malaysia is divided into Western, Central, and Eastern tectonostratigraphic belts based on major geological and geophysical phenomena. The Kinta Limestone is a Paleozoic succession located within the Western Belt. Due to structural and tectonothermal complexity, the sedimentological and paleontological works in these carbonates have proven to be problematic unless combined with geochemical approach. Thus, the current study has integrated stratigraphical, sedimentological, and geochemical studies to assess the lithofacies variations and to interpret the depositional environments. An intensive fieldwork has been carried out in order to assess the extent of metamorphism and to locate the less altered sections for further studies. Three boreholes have been drilled on N-S transect of the Kinta Valley recovering a 360 m core. The core description, the mineralogical analysis, and the geochemical analyses including major and trace elements and organic carbon contents have allowed for a significant advancement of the knowledge existing on this basin. The obtained results have indicated that the Kinta Limestone is chiefly composed of carbonate mudstones, siltstones, shales, and minor cherty units. It preserves the main sedimentary features from metamorphism, especially in the northern part of the Kinta Valley. The detrital siliciclastic debris is minimum in the limestones. The overall dominance of fine-grained textures, the lacking of detrital siliciclastic deposits, presence of bedded cherts, and high organic carbon content outlined by geochemistry and the occurrence of uncommon benthic fauna have suggested the deposition in a slope environment with low energy and low oxygen content. The lithological changes from carbonate to siliciclastic deposits have outlined the occurrence of sea level fluctuations in the Paleozoic. The various analyses combined with chemostratigraphy, an independent of type locality and stratotype, enable to interpret the depositional environment of the Kinta Limestone. Thus, it can be useful to correlate to other formations in or similar types of basins in the southeast Asia.

Keywords: Kinta Limestone, Paleozoic, chemostratigraphy, Kinta Valley

1. Introduction

This chapter contains primary research data from a research project conducted on the stratigraphy and sedimentology of the Kinta Limestone. It has attempted to incorporate reviews on the stratigraphy and tectonic evolution of the Peninsular. It is widely accepted that the Kinta Limestone is one of the massive carbonate deposits which has longer temporal extension of Silurian to Permian and spatially covers the central part of the Western Belt of Peninsular Malaysia. In this contribution we have attempted to use relict primary sedimentary textures combined with geochemical signatures of cored and outcropped sections to interpret the environment of deposition and prevailed conditions during the deposition of the Kinta Limestone.

1.1 Geologic and stratigraphic setting of the Southeast Asia

Peninsular Malaysia is located in the southern-most tip of Asia Mainland (**Figure 1A** and **B**). It covers a total area of 130,268 km^2 and forms part of the Sundaland with the number of smaller islands emerging from the shallow seas. It is elongated in NNW-SSE direction and characterized by a dense network of streams

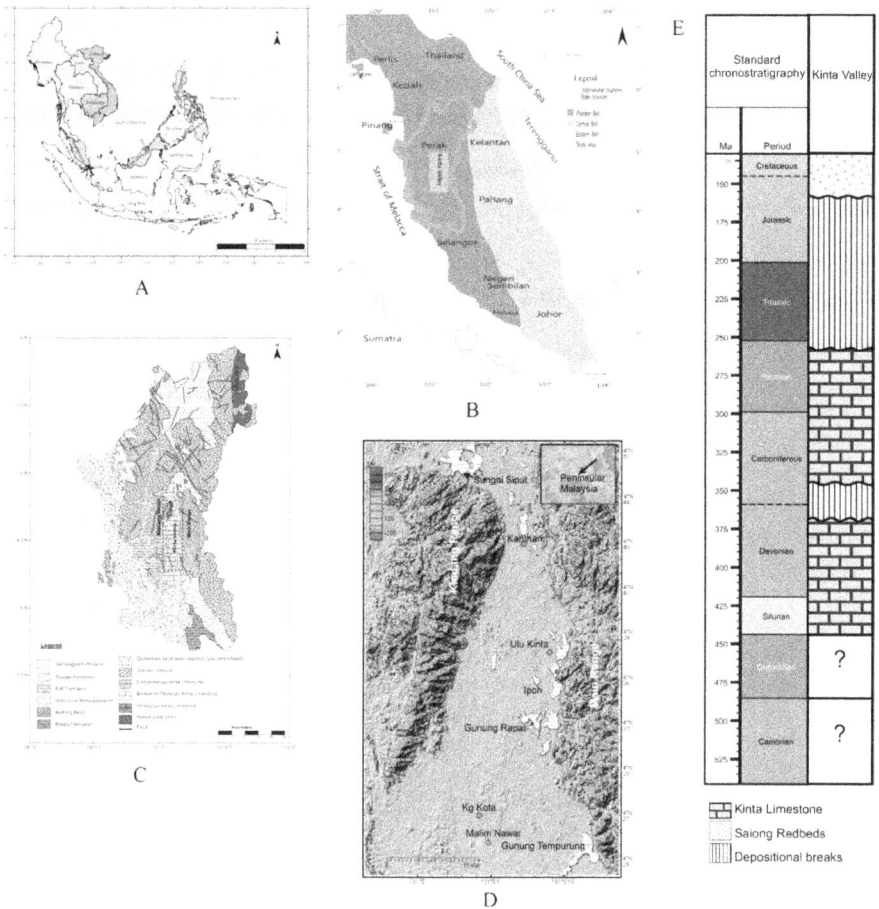

Figure 1.
Location map of the study area with respect to Southeast Asia regional map (A), tectonostratigraphic zonations of the Peninsular Malaysia (B), Geological map of the the Perak State (C), digital elevation model (D) of the the Kinta Valley area indicating the distribution of outcropped Kinta Limestone hills, and the composite stratigraphy of the Kinta Valley area (E) modified from [53]. Note that drilling locations are in the north (Sungai Siput) and in the south (Malim Nawar).

and rivers [1], which exposes older rocks, particularly in the Northwestern Domain of the Peninsula and the Kinta Valley.

Thick and widely spread carbonate deposition is common during the global sea level highstand [2–7]. During the Paleozoic, the global sea level was rising to a larger scale [8–10], during which the paleogeographic position of the Sibumasu was tropical to subtropical [11]. Most of the carbonate depositions were formed in tropical to subtropical latitudes, and some of the Phanerozoic carbonates even formed in lower latitudes [12]. The growth and the distribution of the carbonate successions in the south-eastern Asia region would not be an exception; hence, there should be a link with the paleogeographic evolution of the region during the Paleozoic times. Particularly, the limestone in Peninsular Malaysia might be strongly associated to the tectonic evolution of the peninsula during the opening-closure of the Paleo-Tethys. In this perspective, Peninsular Malaysia is made of two continental blocks, the Gondwana-derived Sibumasu and Indochina terranes, which come together in the Late Triassic. According to Metcalfe [13], the boundary between the Sibumasu and Indochina terranes was marked by the closure of the Paleo-Tethys Ocean and was evident in the Bentong-Raub Suture Zone, representing a segment of the main Devonian-Middle Triassic Paleo-Tethys Ocean. The present-day Peninsular Malaysia is formed from these two terranes and is divided into three main different tectonostratigraphic zones, having a characteristic stratigraphy related to their tectonic history [14, 15]. This research has been carried out in the Western Belt, which includes the Northwestern Domain, which was part of the north-western Australian Gondwanaland during the Late Paleozoic. During the Late Permian and during the Middle to Late Cretaceous, two compressional events were reported in the Peninsular Malaysia [16]. Tectonothermal events have left confusing structures of folding to secondary sedimentary structures such as slumps. The tectonic uplift coupled with isostatic movements has caused considerable loss of sediment thickness on the Pre-Permian and Permian rocks of the peninsula, due to erosion, resulting in the deposition of younger sequences such as the Semanggol Formation [11]. However, the stratigraphic boundaries of the major Paleozoic formations of the Western Belt of Peninsular Malaysia are not clearly known due to the complex structural, thermal events, and lacking accessible stratigraphic units. This is even more complicated in the Paleozoic carbonates of the Kinta Valley, which are represented by scattered protruding limestone tower hills. Major portion of these limestone hills are covered by thick Quaternary deposits and surrounded by the massive granite bodies in the east and west marked by the elevated regions on the map (**Figure 1C** and **D**).

1.2 Historical development on the stratigraphy of the Kinta Valley

Peninsular Malaysia can be subdivided into three north-south trending belts (**Figure 1B**) with characteristic stratigraphy, structure, magmatism, geophysical signatures, and evolutional paths [11, 14, 17]. The stratigraphic classification for the Paleozoic successions of the peninsula comprises 42 formations [14, 18], and few of them have been revised to establish new lithostratigraphic units [14, 19]. For instance, the lower Paleozoic rocks are found in the western part of the peninsula, whereas upper Paleozoic successions are common in all the three stratigraphic belts of the peninsula [14, 18]. The Kinta Limestone (**Figure 1E**) has been assigned to varying geological ages depending on its geographic locations ([20] and references therein) and other authors cited in [14, 18, 21, 22]. Among the earliest to share their findings on the geology of Perak State were Errington and Wood ([23] and references therein). The first attempt to establish the lithologic succession of the Kinta Valley was made by De Morgan [24] followed by Wray [25], who classified the lithology of the Kinta Valley into four rock types in which the crystalline limestone was overlain by a large series of beds comprising gneiss, quartzite, schist, and sandstone as cited in Ingham and Bradford [23].

In 1903, another view from Collet had emerged where he believed that the limestone is younger than the other sedimentary beds ([23] and reference therein). However, Scrivenor [26] has published a study with the crystalline limestone as the oldest unit compared to the Mesozoic granite, phyllites, and quartzite. More specifically, Scrivenor [27] considered the limestone (carboniferous) to be older than the phyllites and quartzites in which he addressed the clays and boulder clays as "Younger Gondwana Rocks." However, in 1917, Jones disagreed with the view of Scrivenor [27] specifically on the source and nature of the boulder clays and pro-posed a sequence in which the limestone belongs to Permo-carboniferous periods [23]. Thus, in Jones' lithological succession, the crystallized limestone was stated to be underlying the schist, phyllites, quartzites, indurated shales, granites, and alluvial deposits, which was consistent with Scrivenor [27]. It is also mentioned in Ingham and Bradford [23] that Cameron [28] and Cameron [29] had disagreed and proposed a new geological succession. This is simply to highlight how the estab-lished stratigraphy of the peninsular, particularly the Kinta Valley, was complex and debated among researchers for a century and half.

In the study area, the chemical characters of the rocks have been used to infer their depositional conditions and to develop correlation schemes. Redox-sensitive elements, coupled with major and trace element analyses have been carried out to assess the Kinta Limestone at the level of depth that this project has reached.

2. Stratigraphy and geochemical analyses

2.1 Stratigraphy

In this chapter, we present our findings on all the accessible outcrops along a north-south transect of the Kinta Valley and the stratigraphic data from three shallow boreholes drilled into the Kinta Limestone. The study has included all major outcrops from Sungai Siput through to Malim Nawar (**Figure 1D**). A detailed fieldwork has allowed us to choose selected outcrops near to the Sungai Siput, enabling to infer the depositional conditions of the Kinta Limestone. These hills are outliers surrounded by siliciclastic deposits, relatively less effected by metamor-phism, and have exceptionally preserved primary sedimentary features [30]. Two boreholes, SGS-01 and SGS-02, were drilled and retrieved a total of 126.98 m of cores. These boreholes were approximately drilled to a lateral distance of 1 km from each other. A third borehole, MNR-03, was drilled further to the south of the Kinta Valley in Malim Nawar (**Figure 1D**) and retrieved the deepest core, at 232.82 m vertical depth. This is an area where most of the fossiliferous surface limestone sites were reported in the literature. The three boreholes give a total of ~360 m of core recovery, which enabled detailed lithofacies and micropaleontological studies of the Kinta Limestone.

The lithofacies in the northern part of the Kinta Valley is mainly dominated by dark to black carbonate mudstone with black shale beds and siltstone intervals, particularly at the base of the boreholes SGS-01 and SGS-02. The southern section of the Kinta Limestone contains calcitic limestone with minor clastic intervals. The southern section has relatively coarser-grained texture than the northern section lithofacies of the Kinta Limestone. It has also been revealed that the Kinta Limestone has older sections in the present north and younger sections in the pres-ent south of the Kinta Valley [31].

2.2 Geochemical analysis

2.2.1 Inorganic geochemistry

In this research, X-ray diffraction (XRD) and X-ray fluoresces (XRF) were used for mineralogical, elemental and oxide analysis of selected samples. X-ray diffraction (XRD) is one of the basic tools in the geochemical analyses of rock samples of limestones and mudrocks [32]. The samples were prepared in the Material Laboratory of Universiti Teknologi PETRONAS. The rock samples were pulverized to grain-size less than 63 μm in order to obtain a well-homogenized powder, according to the methods and recommendations of Tucker [32]. The powder was dried in an electric oven at 49.3°C for about 24 hours to remove the moisture content. Once the samples were dried, they have been analyzed using the XRD machine, in which the diffraction patterns have been interpreted using dedicated software and databases [33, 34]. The analyses and interpretations of the XRD results were based on the procedures stated by Tucker [32] and Dong [34].

There is an increasing trend in the application of XRF techniques in the characterization of carbonate rocks, particularly for elemental chemostratigraphy of carbonates and useful tool to study the stratigraphic successions. XRF is the preferred technique for the analysis of major and minor elements, such as Si, Al, Mg, Ca, K, Na, Ti, S, and P in siliciclastic rocks and also for trace elements such as Pb, Z, Cd, Cr, and Mn. An energy-dispersive XRF system was applied in this research since it has the ability to measure all elements over a spectrum of wavelength. A total of 12 representative outcrop samples were collected from Kanthan, Malim Nawar, Ulu Kinta, and around Kampar areas (**Figure 1D**) for geochemical analysis using XRF. Forty-four borehole samples were analyzed for elements and oxides. These samples were collected to represent the major lithofacies variations noted during core description. The sample distribution is indicated in **Table 1**.

The samples were mainly carbonate rocks except four siliciclastic rock samples, which were included from the northern part of the Kinta Valley. The results are plotted for the samples, which have values greater than zero based on the detection limits of the testing tools.

2.2.2 Organic geochemistry

In a broader sense, organic geochemistry deals with the fate of carbon, in all its variety of chemical forms in the earth system [35]. Many organic matters can be incorporated into sedimentary rocks and preserved for millions of years. Kennedy et al. [36] showed that the preservation of organic carbon is most commonly controlled by oceanographic regulation of bottom-water oxygenation and/or biological productivity. This relationship could be used to evaluate the organic content of sedimentary rocks and the records on paleowater conditions such as energy level, water depth, rate of deposition, and other related paleo-oceanographic parameters.

Sample source	Quantity
Outcrop	12
SGS-01	10
SGS-02	14
MNR-03	20

Table 1.
Number of samples for XRF analysis taken from the respective localities.

Total organic carbon (TOC) content and pyrolysis analysis using Source Rock Analyzer were carried for a total of 60 samples. These analyses were selected in order to support both the interpretation of the depositional environment (the vitrinite reflectance, distribution of TOC with grain size); the paleothermal evaluation (Tmax that is compared with the temperature from the CAI). The TOC measurement provides clue on the paleodepositional conditions such as oxic and anoxic, indicating also the rate of deposition and level of surface organic productivity.

2.3 Chemostratigraphy

Chemostratigraphy is application of sedimentary geochemistry to stratigraphy. It is highly linked to lithostratigraphy since the major lithologic differences are connected to changes in the importance or balance between different geochemical reservoirs. It is based on the theory that the chemical and physical composition of sea water varies through geological time. These changes have been recorded by the chemical composition of the sediments (by major and trace element distribution) and by the isotopic ratios of particular elements [37, 38]. The seawater composition is considered homogenous for a certain geological time, since the mixing rate of seawater is relatively short; therefore, geochemical variations have global implications [37]. The collected outcrop and core samples were analyzed for major and trace elements. The main interest in the analytical techniques was to characterize the rock units in terms of elemental concentration and ratios and to uncover any systematic trends. The use of changes in elemental composition of sediments of the outcrop and borehole samples has enabled to establish a chemostratigraphic zonation. Thus, geochemical analysis is used in conjunction with petrographic data in the selection of samples for the establishment of chemostratigraphy and interpretation of the paleoenvironment. Chemical stratigraphy is a study on the variation of chemistry within the sediments or sedimentary sequences for stratigraphic correlations [39]. It is a tool that uses the changes in elemental composition of sediments to characterize sedimentary sequences and to spatially extend this characterization between outcropping sections or wells to form a chemostratigraphic correlation. This characterization enables the identification of chemostratigraphic packages and units on the basis of elemental concentrations, ratios, and their systematic trends. Chemostratigraphy is a relatively young subdiscipline in the stratigraphy, and it is also a new attempt to apply it on the Kinta Limestone.

3. Results

3.1 Lithostratigraphy of the Kinta Limestone

The surface lithology of the Kinta Valley is dominated by the granites, overlain by Quaternary-Recent deposits (**Figure 1C**). Paleozoic deposits occur as isolated exposures. Three types of sedimentary lithologies namely, limestone lithofacies followed by siltstone and thin black shale beds in the order of decreasing proportions, were identified based on outcrop, core, and petrographic studies. Rarely bedded chert units also occurred in specific localities, for example, in the Sungai Siput section. The Sungai Siput section, located in the northern part of the Kinta Valley, is dominated by fine-grained, light-to-dark gray to black, thinly laminated carbonate mudstone. It shows a sharp contact with the light-gray and dark-colored carbonate mudstone, cherts, and the black shale lithofacies. These lithofacies types at place exhibit slump structures. The section in Sungai Siput site is less affected by dissolution; however, it is highly compacted and stylolitized if compared with the other sections of the valley. The samples from the dark gray intervals of the Sungai Siput sections were found to be richer in phosphatic microfossils in contrary to the other lithofacies [31]. When compared with the very thick carbonate mudstone beds (many tens of meters thick), relatively thin black shale (2 m thick) in SGS-01 borehole was typical of this unit. The light brownish-gray, fine-grained muddy siltstone lithofacies occurs in the Sungai Siput sections of SGS-01 and SGS-02. This muddy siltstone underlying the carbonate mudstone is a dominant lithofacies intercepted in boreholes SGS-01 and SGS-02, but not encountered in borehole MNR-03 in the south of the Kinta Valley. The siltstone is highly affected by fractures and compositionally dominated by quartz. An outcrop section located further south-east of the Sungai Siput (Kanthan) is characterized by light gray, fine-grained, subhorizontally bedded thin to thick intercalations of carbonate mudstones. It contained dark-colored fine-grained material on the contact planes of the bedding surfaces. Minor whitish to white pinkish colored lithofacies was observed in the section near the granite contacts. The degree of lithofacies heterogeneity is less than the Sungai Siput section. Syn-depositional sedimentary structures, including slumps and thin bedding with sharp contact surfaces, were ubiquitous in the intervals dominated by the black shale interbeds and dark gray carbonate mudstone of the Sungai Siput section. Other sections, located in the eastern sector of the valley, are found to be entirely dominated by metamorphosed units, where most of the sedimentary features have been obliterated. There is no observable lithological variation on the metamorphosed carbonate hills in the western foothill of the Main Range Granite such as Gunung Rapat (**Figure 1C**). Rather, coarser and homogeneously calcitic marble occurs in these areas. Along the contact of planes of the metamorphosed carbonates, pyrolusite mineralization was noted. Faults and joints are also observed with well-developed dissolution cavities and large-scale caves (e.g., Kek Lok Tong, Gua Tempurung). Three sets of sealed fracture networks filled with calcite cement also occurred. The degree of surficial weathering has impacted the lithology variably with respect to geographic locations in the Kinta Valley. For instance, the dissolution is intensive near to the granitic intrusions, where metamorphism is higher and in the southern part of the valley, where subsurface pinnacles were found.

The siliciclastic interval of the cored sections of the Kinta Limestone shows higher gamma-ray (GR) values, whereas the limestone intervals are characterized by low values and smooth gamma ray log curves. The density log of the section is also in agreement with GR logs in such a way that the higher GR values are associated with high density. The SP log was not diagnostic, except for its indication of the occurrence of a porous siliciclastic interval at the base of the SGS-02 borehole. Caliper log detected the lithological contacts between the limestone and the siltstone intervals, which was denoted by caving in the siliciclastic intervals, while the limestone portrayed a fairly smooth borehole environment, as the limestone is relatively compact.

3.2 Chemostratigraphy of the Kinta Limestone

The samples from borehole SGS-01 were collected based on the core description from top to bottom of the borehole at irregular intervals. The result shows the four elements (K, Si, Ni, and Al) are very low in concentration in the pure limestone sections. K, Si, Ni, and Al are present in high concentrations in the shale beds found from 9.87 to 13.43 m depth (**Figure 2**). These elements are also present in high concentrations from 40 down to 65 m depth. Moreover, all the four elements have shown slight variations along the vertical profile of the borehole with respect to lithologies (**Figure 3**).

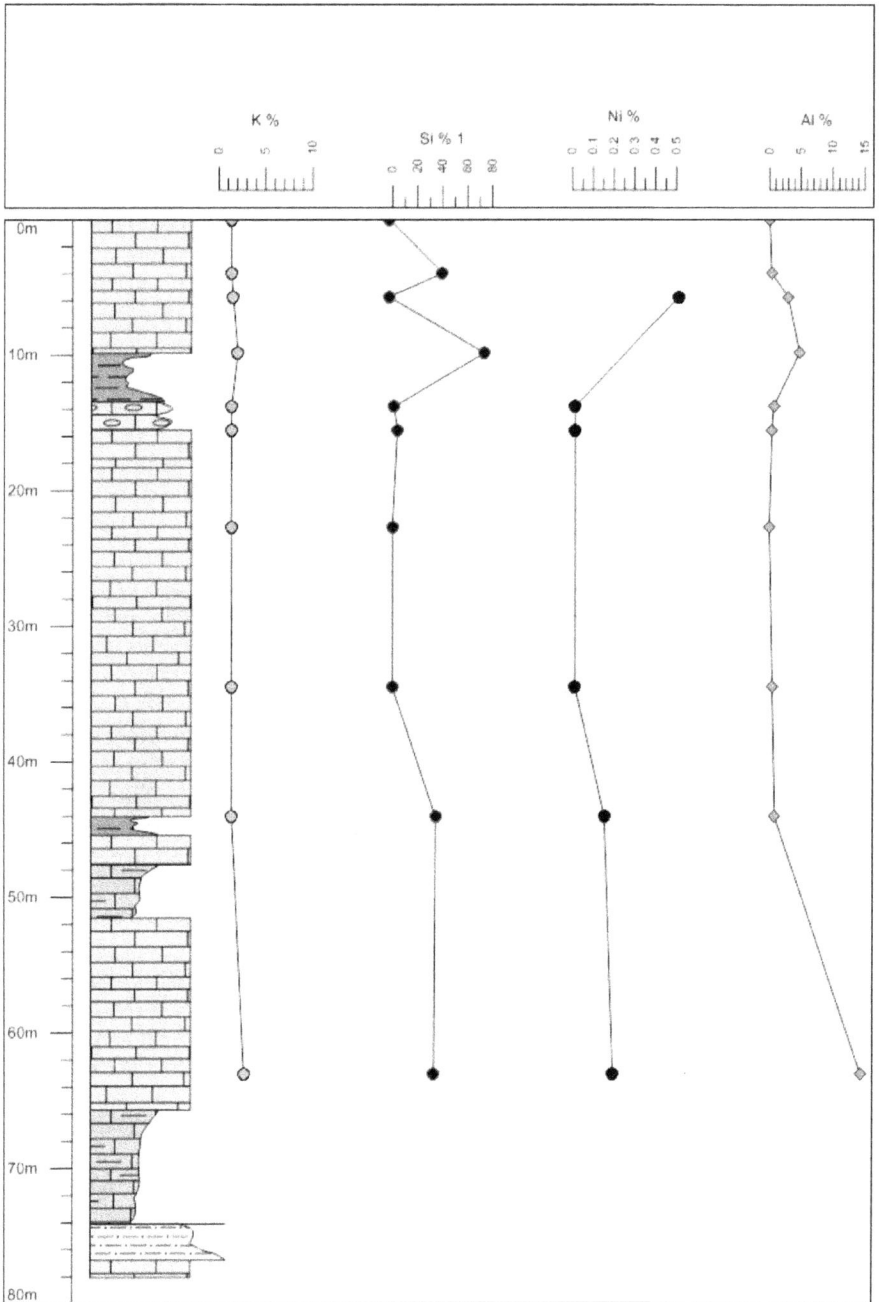

Figure 2.
Chemostratigraphic log of borehole SGS-01. Note that the variation of these K, Si, Ni, and Al along the vertical profile of the borehole.

Ca, Mg, Mn, Sr, Zr, and their ratios have shown a significant variation with depth and a wiggling pattern along the vertical section of the borehole (**Figure 4**). The Ca content of the carbonates is high, whereas Mg is low (less than 1 wt.%). Ca and Mg are among the major elements in the carbonate mineralogy, while Mn and Sr are minor but provide a clue of the postdepositional history of carbonate rocks. Si is lowest for carbonates but it is highest for the siliciclastic deposits of the

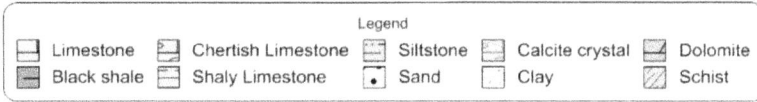

Figure 3.
Lithology legends for the lithostratigraphic logs in the boreholes. Note this legend is used throughout this chapter.

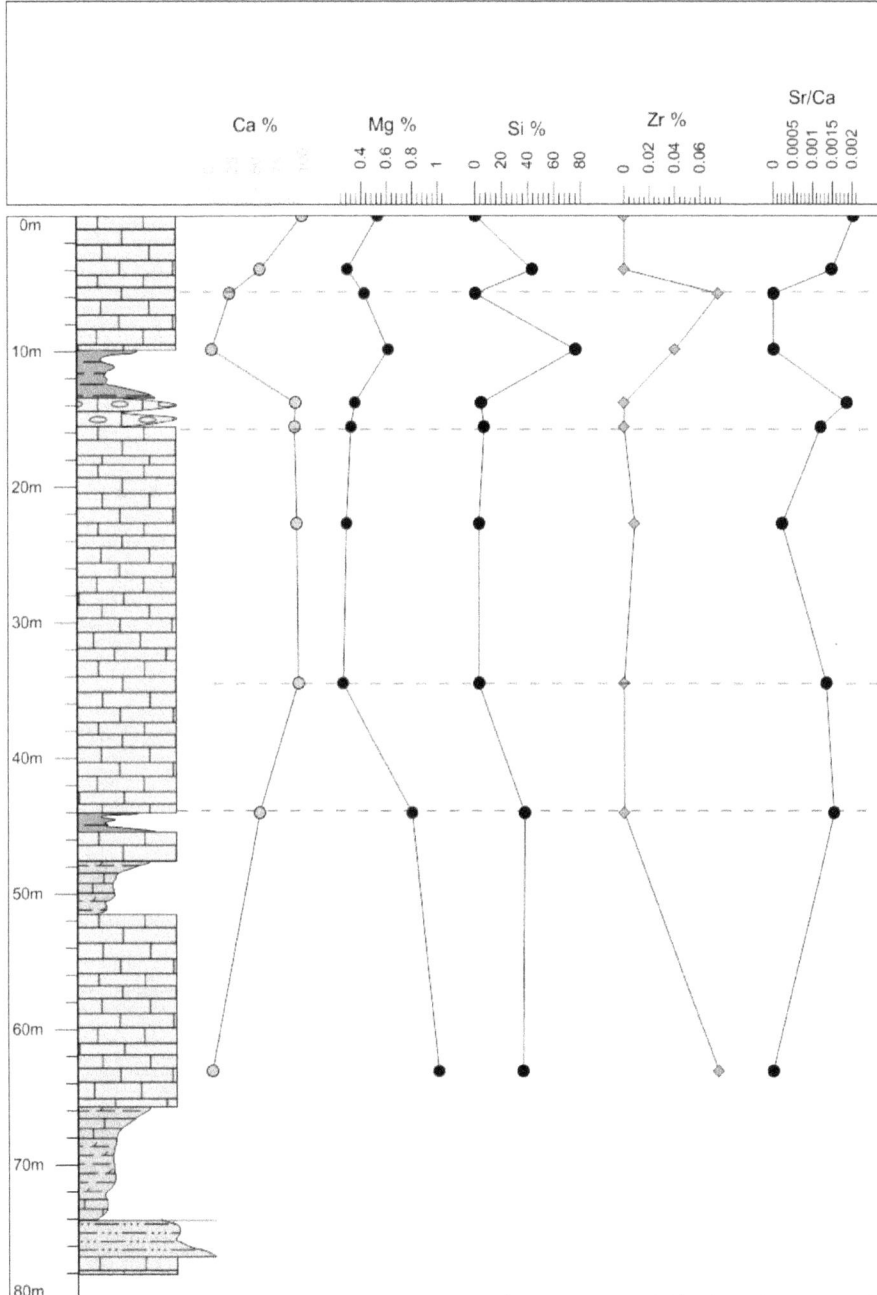

Figure 4.
Chemostratigraphy of the SGS-01 based on elemental concentrations and ratios. Note that there are a few chemical packages, which are characterized with similar chemical signatures for the displayed element and ratio data set.

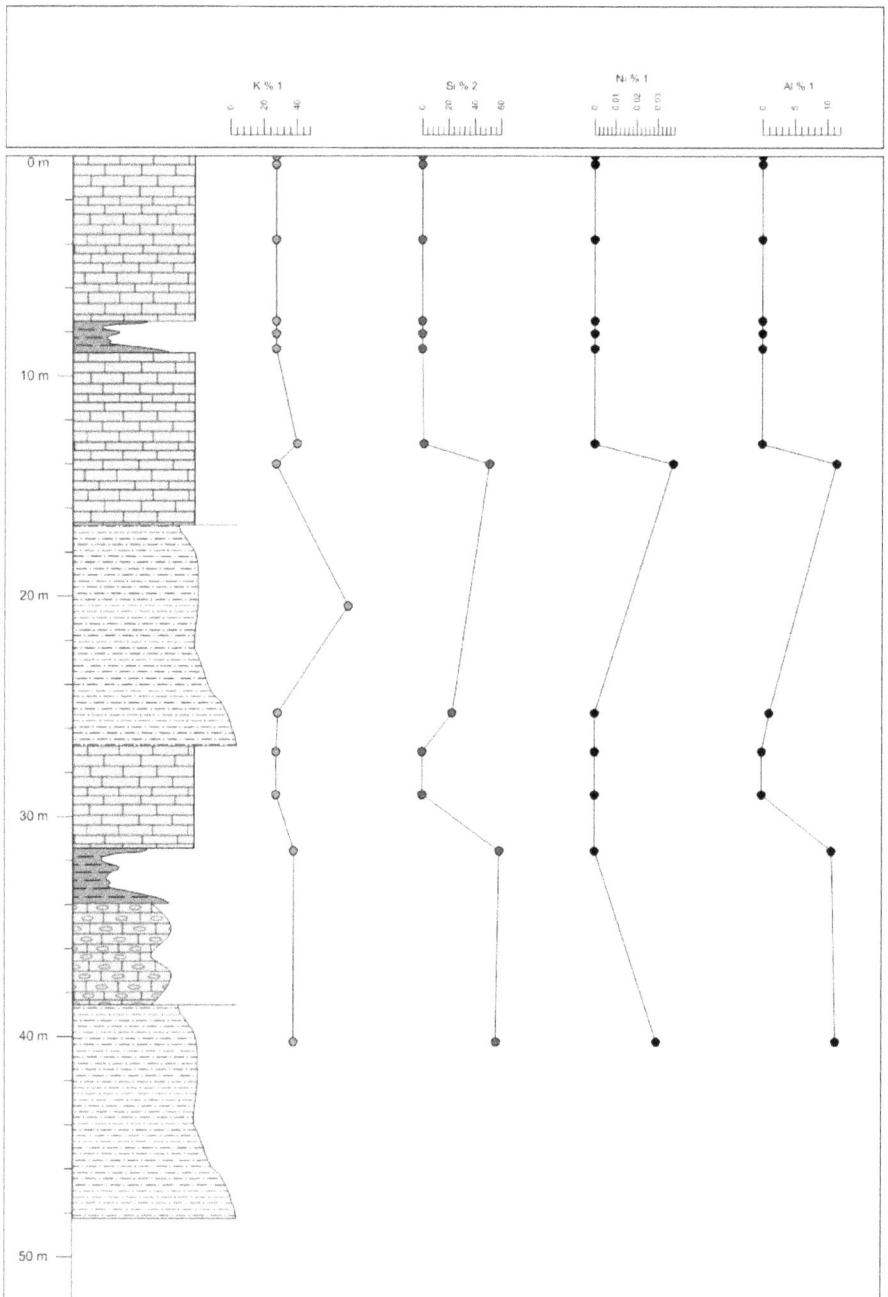

Figure 5.
Chemostratigraphic log of the borehole SGS-02. Note that the slight correlation of the elemental concentrations of the data set.

borehole. The Si content shows similar trends to the Zr content of the black shale intervals. Total carbon content (TOC) analysis of the samples shows that the black shale intervals have the highest concentration (4.93 wt.%) and lower TOC for the other lithofacies. The correlation with the lithological variations is marked, especially at the top of the SGS-01 section.

Although a slight difference in the lithological arrangement, the elemental concentration of K, Si, Ni, and Al in the borehole SGS-02, has been noted, it has shown similar trend as in borehole SGS-01. The concentration of K is higher in the siliciclastic lithofacies such as the shale and the siliceous carbonates (**Figure 5**). Si, Al, and Ni also followed a similar pattern in the siliciclastic intervals. These elements

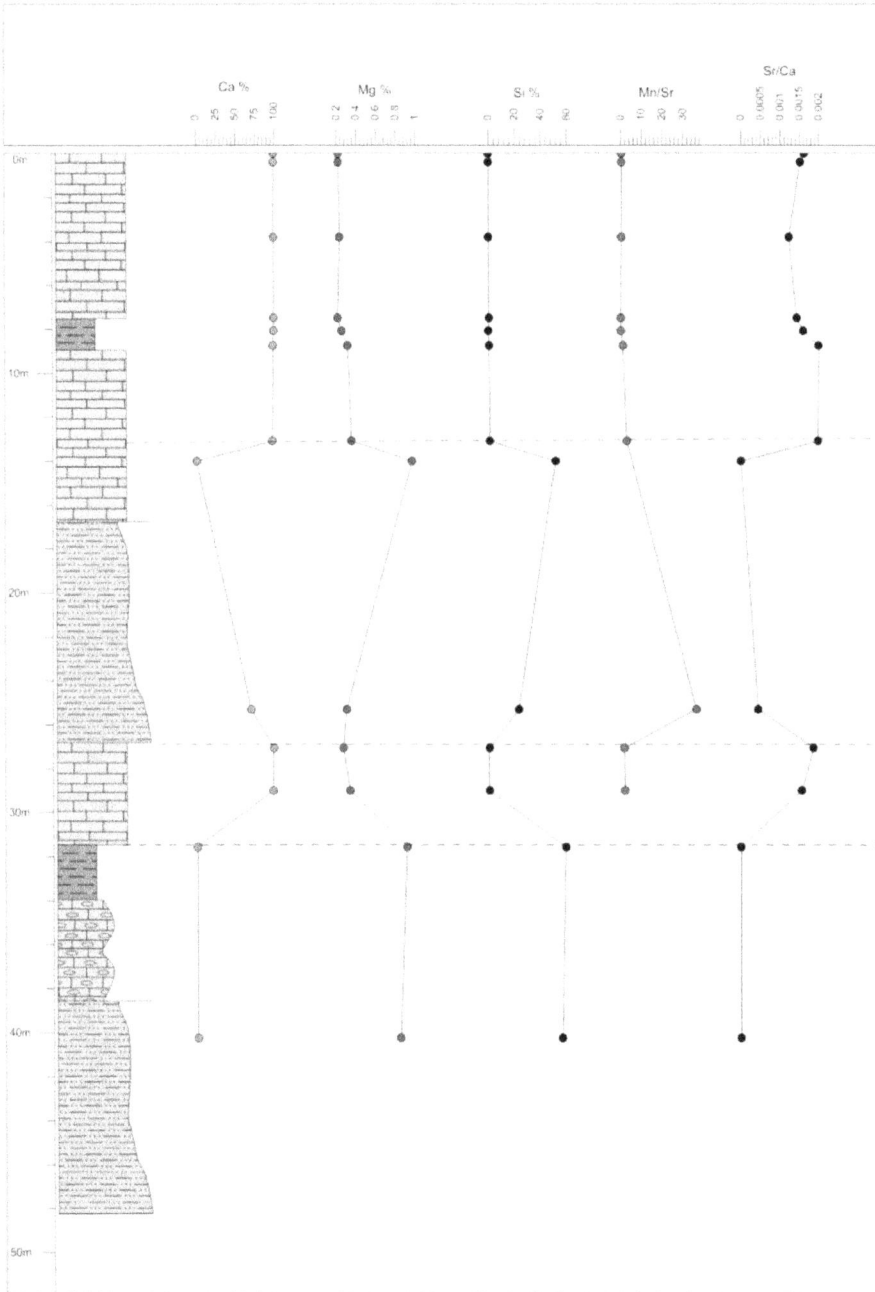

Figure 6.
Chemostratigraphy of SGS-02 based on elemental ratios. Note that there are four chemical packages, which showed different chemical signatures.

are either not present or found in very low concentrations in the purely limestone intervals. These variations reflect the lithofacies variation along the vertical profile of the borehole. These chemical signatures may have some implication for the depositional environment and diagenesis history of the section.

A plot of Ca, Mg, Si, Mn/Sr, and Sr/Ca with respect to depth is shown in **Figure 6**. The chemical signature has shown a shift at 13.16 m in all the elements and their

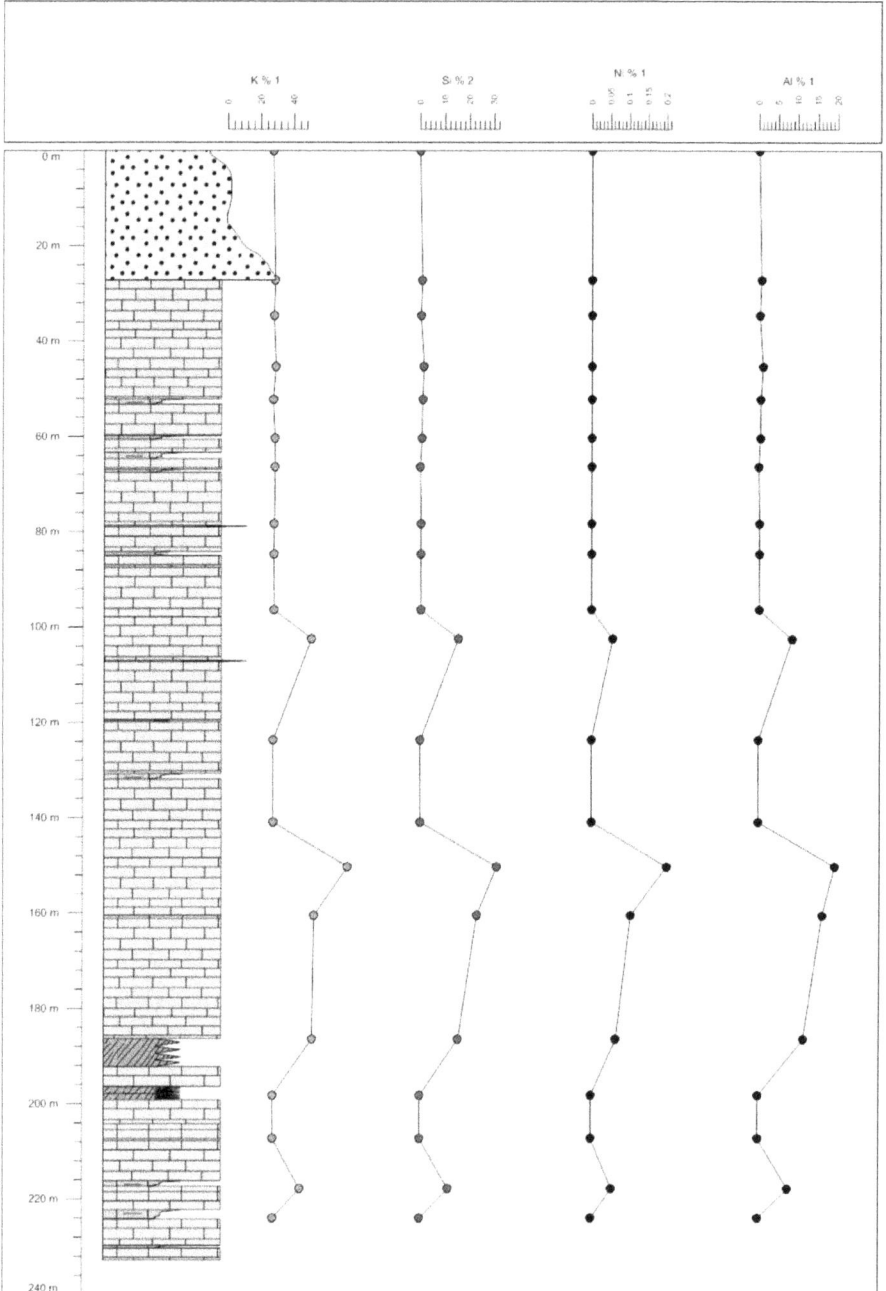

Figure 7.
Chemostratigraphic log of borehole MNR-03. Note the sharp increment of K, Si, Ni, and Al concentrations from 148 to 192 m depth. This depth was also noted having some indication of aerial exposure.

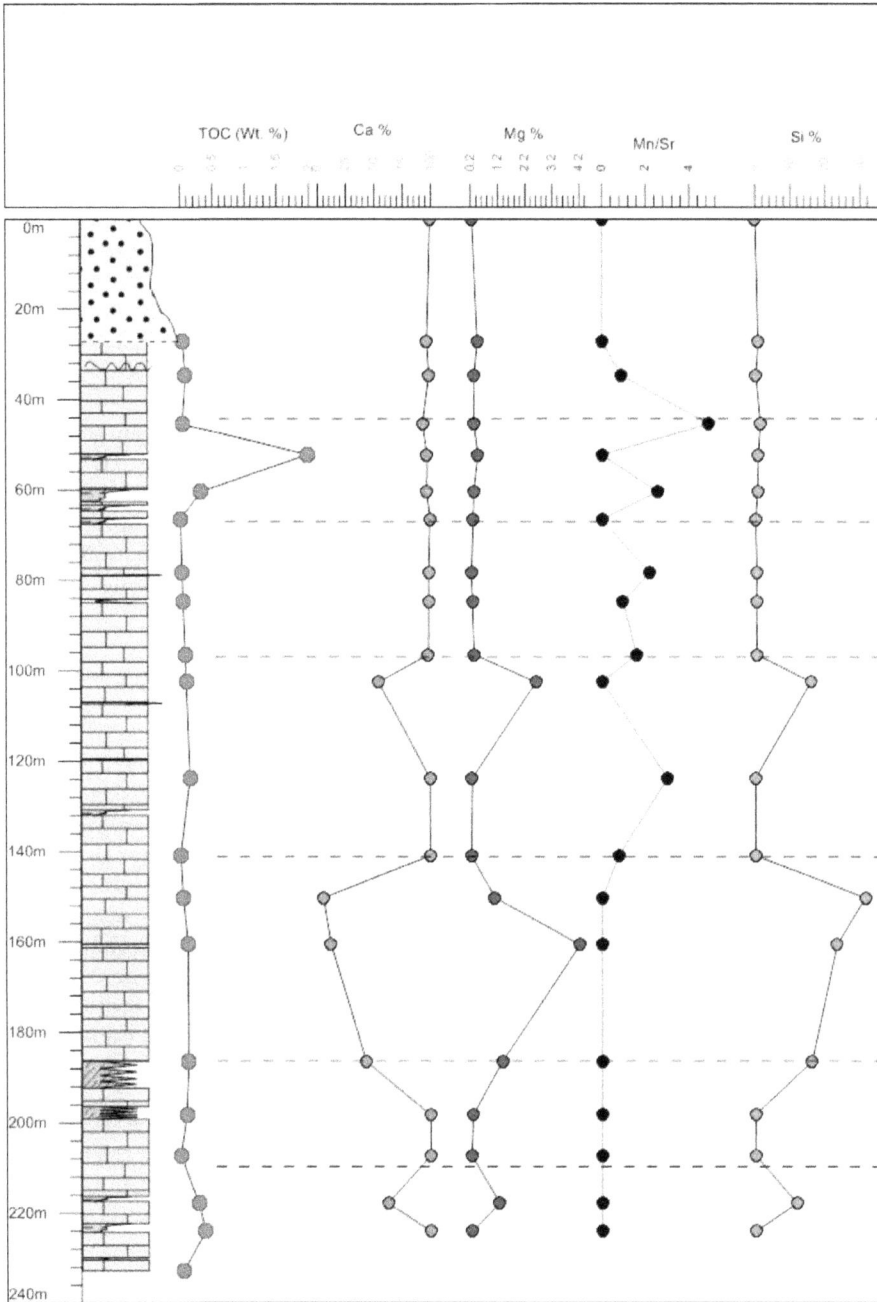

Figure 8.
Chemostratigraphy of the borehole MNR-03 based on elemental concentrations and ratios. Note the pattern of variation in the chemical signature of Ca, Mg, and Si from 148 m downhole to 192 m, where Ca showed decreasing trend, and Mg and Si are increasing until 192 m depth.

ratios. Another significant change of the chemical signature of the lithofacies has been observed to a depth of 26.75 m. Another equally important variation in the elemental signatures is shown at depths of 31 m. Ca is the highest concentration in the carbonate mudstone lithofacies, with little Mg and almost zero Si. These geochemical signatures disclose a direct correlation with the lithological variations.

The TOC is highest in the shale intervals at 7.5–9 m and remains flat in the other lithologies. It can be noted that there may be a possible link with the depositional environments and with chronology of the chemical signatures.

A thorough analysis of the elemental composition of the section from the Kampar area (borehole MNR-03) has showed a particular pattern of K, Si, Ni, and Al. A unique signature with depth from top toward the bottom was depicted by the elemental concentration of the selected elements (**Figure 7**). The limestone in the topmost 100 m of the borehole is almost pure. It contains a very low concentration of K, Si, Ni, and Al. Minute peaks of the four elements occurred from 96 to 115 m. A major increase in the elemental concentration of the four elements occurred from 148 to 192 m. This is then followed by a section of pure limestones with a small shift of the four elements until a depth of 217 m. These geochemical signatures are relevant in order to discuss the depositional and postdepositional events.

For the southern part of the Kinta Limestone, the Ca, Mg, and Si elemental concentration showed some enrichment downhole. Ca, Mg, and Si remain some-how constant until the depth of 96 m. From 96 to 148 m, there is a slight variation in all the elemental and ratios. The trend of Ca and Mg is in inverse relationship. From 148 m downhole showed main decrease in Ca and increase of the others, while the Mn/Sr ratio remains constant. The TOC is generally low and only showed high peak at 45–60 m depth (**Figure 8**). These variations may provide some clues about depositional and postdepositional processes on the studied sections.

4. Discussion

The lithofacies distribution in the Kinta Valley depicts important clues for the interpretation of depositional environments. For instance, the lithofacies from the Sungai Siput sections of the Kinta Limestone is mainly dominated by fine-grained carbonate mudstones. In addition to the carbonate mudstones, siltstones, black shales, and chert beds have been described from the northern part of the Kinta Valley. These lithofacies have textural attributes and field distributions, suggesting low-energy depositional environment. To the east of the northern Kinta Valley, the presence of schists was noted, thus providing some implication to infer possible paleodepocenter of the Kinta Limestone. Thus, the textural features of the lithofacies encountered in the present-day northern and north-eastern part of the Kinta Valley imply that the lithofacies reflect a certain type of paleobasinal configuration during their deposition.

The lithofacies colors of the Kinta Limestone varied from light gray to dark and black in the northern part of the valley. This has been used as one of the clues that may indicate the deposition of this type of lithofacies in a condition where and when it is conducive for the incorporation of organic matter. It has been found to be rich in TOC and interpreted that the color variation was partly controlled by the organic content of the rocks. This may indicate that the limestone was deposited in an environment that prohibited oxidation of the organic contents that were incorporated into the sediments during deposition. Conditions for this type of preservation are common either in a restricted lagoon or in a deeper marine setting where less oxygen is available, which led to the survival of limited scavengers and decomposers. Therefore, the best fit for this type of depositional setting would be deeper part of a slope in the northern end of the Kinta Valley; and the slope was shallower toward the south of the Kinta Valley. The presence of shell fragments in the Sungai Siput (N) limestone has an indication that the Kinta Limestone was deposited at a depth shallower than the CCD, since the dissolution of shell fragments made of calcium carbonate is expected to increase with increase in hydrostatic pressure and decrease in temperature with increased water depth. Thus, the origin of the Kinta Limestone should be biogenic with minimal chemogenic input. This is consistent with the observation in the modern carbonates, which are rich in skeletal components of marine organisms; and this was believed to be true in the Phanerozoic time as well [40]. In summary, the texture, color, TOC, and lithofacies analysis of the Kinta Limestone indicates low energy, calm, and varying paleodepositional settings. It was found that most of the lithofacies textures, TOC, biofauna, and color showed that the Devonian-Carboniferous interval of the Kinta Limestone was deposited in a deeper environment, perhaps with limited oxygen supply.

Sedimentary structures could be depositional, which are formed during deposition, or postdepositional, which are formed after deposition of the sediments during the burial stage. The depositional structures are known to be primary, and the postdepositional structures are secondary. The primary structures in the Kinta Limestone are the thinly bedded, dark gray to black carbonate mudstones in the Sungai Siput (N) and the chert laminae. The contact planes of the beddings are sharp, indicating a calm and less disturbed depositional environment. The slumps may signify that the Sungai Siput (N) section might be deposited in a sloping localized bathymetric variation with slight instability right after deposition. Structures that are similar to slump but difficult to confirm are also seen in the Kek Lok Tong areas. These structures are not used for the above inferred interpretation due to the fact that the lithofacies in this area is highly affected by contact metamorphism. This is reasonable as this area is in contact or in a short distance from the Main Range Granite, which is expected to have greater impact on the sedimentary rocks due to its huge source of heat for the metamorphism event. In the geological records, slumps commonly occurred in two locations, high up in the slope and lower down of the slope profile due to the susceptibility of slope facies to downslope creep [41]. The origin and use of the slump structures as indicators of paleoslope have been documented in many works [42–48]; thus, there is a consensus on their use for inferring the paleobathymetry and possible triggering means of the movement of the sediment on the paleoslope. The triggering mechanism could be failure of slope, which is driven by gravity or involvement of seismicity [44, 49]. Such kind of sedimentary structures are known as soft-sediment deformation structures, which provide keys in understanding the depositional history of sedimentary rocks. Soft-rock deformations are documented that they are created either during deposition or shortly after burial [50]. Slumps are common in mudstone, limestone, deposited on a steep slope. They are commonly faulted and folded. They can be indicative of fast deposition rate [51].

Studies on the Devonian-Carboniferous successions from the Northwestern Domain of Peninsular Malaysia indicated that there were shallow marine deposits with major regressive events during the latest Devonian [19]. This implies that a possible shift of depositional environments from shallow water to a deepening trend toward the northern Perak, where this study was conducted. Cocks et al. [52] indicated that there was a general shallow-water and cratonic areas during the Middle Paleozoic of the Sibumasu Terrane, except for deeper successions in the northern Perak. In addition, Meor and Lee [19] indicated that a deposition of thick red mudstone interbedded with silty sandstone in the Early Carboniferous might contradict the depositional break suggested in the Kinta Limestone [53].

Stylolites are the most visible postdepositional events on the cores and accessible outcrops of the Kinta Limestone. Stylolites are postdepositional irregular discontinuity surfaces, which are believed to have resulted from localized stress-induced dissolution features during burial or tectonic compression. The patterns, density, and size of the amplitude of the stylolites have implications on the amount of siliciclastic impurities and heterogeneity of the carbonates that would increase the chance of nucleation of the stylolite seams [54]. Thus, the stylolites in the Sungai Siput (N) section are more abundant than on any other part of the studied sections in the Kinta Valley. This also goes well with the lithological variation observed in the Sungai Siput section. The presence of the fine-grained siltstone, shale, and carbonate mudstone with bedded chert shows that the stylolites correlate with this heterogeneity. In addition, the stylolite may also show uplift (deeper burial so that there was high overburden) or be originated from compressive pressure. Stylolites are common in fine-grained sedimentary rocks affected by chemical dissolution. Chemical dissolution is also common in pelagic to hemipelagic sediments. This supports the contention that the depositional environment of the Kinta Limestone might be pelagic to hemipelagic.

In summary, all the primary and secondary structures discussed are consistently implied a variation in the paleobathymetry from the present-day north to the present-day south of the Kinta Valley. It was revealed that the northern part of the Kinta Valley was dominated with sedimentary structures, which indicated deposition in a steeply dipping slope, which was probably unstable during the upper Devonian to lower Carboniferous time. It was also learned that the environment to the present-day north of the Kinta Valley was favorable for deposition of fine-grained sediments, which are common in low-energy depositional environments. The bedded chert in the Sungai Siput (N) may also suggest that it was a primary deposition of silica-rich material in a low-energy, relatively deeper depositional environment. Moreover, its association with slump structures may also suggest its function as decollement surface during the soft-sediment deformation of the succession. It is also clear that the depositional environment was above CCD, however, at considerable depth of water.

On a general basis, the studied samples have shown that the variation of the major elements is directly related to the lithofacies types. The carbonates are found to be dominated in Ca and contain little Mg. However, they are found to be poor in silicate-derived elements such as Al, Si, and K. This variation may throw light to the condition and source of the sediments during the deposition of the lithofacies. Thus, the implication of the spatial chemical variation may have also temporal significance and indication that the lithofacies may still maintain primary sedimentary geochemical signatures that could have value in understanding the depositional environment and diagenetic processes.

The geochemical findings from different localities in the Kinta Valley provide clue for the establishment of chemostratigraphic correlations on the outcrops and borehole data sets. A generally trending variation of the chemical and lithological variation from the north to the south of the valley is a typical observation on the analyzed samples. These characteristics were reflected in the elemental ratio and oxides concentrations. Those variations make sense as the rocks are also found to be younger in the south than in the north [31]. Therefore, the variations may have some relation to the compositional variations of the Paleo-Tethys Ocean. The lateral and vertical variations of the chemical signature of the Kinta Limestone and its implication for depositional environment provide clues as to why the silicate-derived elements are rarely detected in the carbonates, which implies that their deposition was not contemporaneous with the carbonates. The absence of silica in the carbonates and absence of carbonate minerals in the siliciclastics may indicate that the limestones are clean of detrital material, and siliciclastics are also clean of carbonate inputs, which implies that there might be two different depositional systems in the upper Devonian to late Carboniferous time in the Paleotethys.

For instance, the concentrations of Mg and Sr can be used to infer the original mineralogical composition of carbonate rocks. The carbonates, which depicted high concentrations of Sr, are inferred to be aragonitic in origin, whereas carbonates, which have high concentration of Mg, are referred to be magnesian calcite in composition, which is a metastable carbonate mineral. These interpretations are related to the size of crystal lattice during a solid solution substitution process. The large unit cell of orthorhombic aragonite crystals can be accommodative to cations larger than Ca such as Sr, Na, Ba, and U, while the smaller unit cell of rhombohedral calcite preferentially incorporates smaller cations such as Mg, Fe, and Mn [55]. These criteria may help to infer the original mineralogy of the carbonates in Kinta Limestone as well. As the carbonates in the north contain a relatively lower Mg than the samples from the south of the Kinta Valley, it may indicate that the original carbonate sediments toward the south of the Kinta Valley were dominated by calcitic carbonates. The Sr concentration in the Kinta Limestone also showed a decreasing trend from north to the south of the valley. This may suggest that there were localized compositional variations during the deposition of the Kinta Limestone in the Paleo-Tethys. Sunagawa et al. [56] have demonstrated that the Sr is preferentially precipitated in aragonite and promotes the nucleation of metastable minerals. The Ca/Sr ratio for the Kinta Limestone also showed spatial variations from north (Kanthan) to the south (Kampar) as the Sr concentration decreased in a similar way. Thereby, this trend may have some relevance to the original mineralogical composition of the carbonates in the Paleo-Tethys too. The Ca/Sr ratio usually is a contribution from the aragonitic carbonates since calcite accommodates very little Sr. Thus, the ratio is important only with regard to warm environments, which are favorable to aragonitic organisms and the inorganic precipitation of aragonite. Therefore, the carbonates in the southern part of the Kinta Valley were calcite dominated, and the carbonates to the north might be aragonitic sediments during deposition.

The Ca/Mg ratios partly reflect temperature-depth-distance from shore and local ecological concentrations of aragonite-secreting algal taxa such as Halimeda [57]. The Kinta Limestone has depicted a systematic trend whereby the Mg concentrations increased toward the south of the Kinta Valley in the Kampar area. The elemental ratios for the selected samples also follow a spatial trend along the valley from north to south, and it may be related to temporal variation as well. If these variations represent the original signature of the Paleo-Tethys, it may contribute to define the depositional environment of the sedimentary and metasedimentary successions in the Kinta Valley. These results are consistent and fairly agreed with the work of Hutchison [58]. During the organic synthesis, magnesium and strontium are incorporated into solid solution of calcite and aragonite, respectively [59]. Gross Ca/Mg ratio of mixed detrital carbonate sediment could reflect temperature-depth-distance from the shoreline [57]. In the case of the Kinta Limestone, it might be possible to suggest that the deposition of the carbonates was not within a timeframe when the basin was getting a huge amount of detrital influxes. Therefore, there might be an episode of alternating depositional periods when the siliciclastic and the carbonates were deposited in a cyclic manner.

In some of the stratigraphic successions, there is a clear physical variation such as color, texture, sedimentary structural differences between strata. Such color differences often originate from variations in the incorporated transition metals during deposition and lithification. Other differences in color may also originate from variations in the organic carbon content of the rock. It now appears that the oceans during early Paleozoic and middle to late Cenozoic time favored precipitation of calcite, probably because of a lower ratio of magnesium to calcium during these times. The ratio of dolomite to calcite is much greater in ancient carbonate rocks than in modern carbonate sediments, presumably because $CaCO_3$ minerals exposed to magnesium-rich interstitial waters during burial and diagenesis were converted to dolomite by replacement [60]. The Kinta Limestone showed a slight difference with the general trend but showed local trends of increasing Mg from north to the south, which may be related to dolomitization or original mineralogy of the carbonates. Comparison in chemical characteristics of the Kinta Valley sedimentary successions in the timeframe of the upper Devonian-lower Carboniferous has shown measurable variations. The measured concentrations and ratios in the elements and oxides would imply that there is no homogenized lithofacies in the Kinta Valley rather the variations provide clues for the Kinta Limestone to improve our understanding on the stratigraphy. This is to convey that the idea of using the relict geochemical signatures of the Kinta Limestone could be a way forward to improve the stratigraphic correlation of the formation with local and/or regionally precisely dated stratotypes.

5. Conclusions

The preserved primary features and textural attributes of the described sections in the Kinta Limestone consistently implied a paleobathymetric variation from the present-day north to the present-day south of the Kinta Valley. It was revealed that the northern part of the Kinta Valley was dominated with sedimentary structures, which suggest deposition in a steeply dipping slope during the upper Devonian to lower Carboniferous time. The environment in the present-day north of the Kinta Valley was favorable for deposition of fine-grained sediments, which are common in low-energy depositional environments. The bedded chert in the Sungai Siput (N) may also suggest that it was a primary deposition of silica-rich material in a low-energy, relatively deeper environment. It is also clear that the depositional environment was above CCD but at considerable depth of water. The chemical stratigraphy of the Kinta Limestone has shown diagnostic characters that enable the establishment of chemical packages, which agreed with the biostratigraphic boundaries for the Kinta Limestone. The sedimentological and geochemical data sets suggest deposition in deep slope with low energy and probably anoxic condition. The various analyses combined with chemostratigraphy, an independent of type locality and stratotype, enable to interpret the depositional environment of the Kinta Limestone. Thus, it can be useful to relate other formations in or similar types of basins in the southeast Asia.

Acknowledgements

It would be our pleasure to acknowledge Universiti Teknologi PETRONAS (UTP) for financial and material support extended to conduct this research. The authors are also grateful for contributions from Professor Dr. Bernard Pierson, Dr. Aaron W. Hunter, Professor Dr. R.P. Major, Professor Micheal Poppelreiter, Ms. Loo Sheau Huey, Mr. Eric Teng Jing Hang, Mr. Abd. Hakim bin Mohd Yusof, and Mr. Rowland Law with his drilling Team in Ujiteknik Geoenviro BHD, UTP laboratory Technologists (Mr. Irwan bin Othman, Mr. Amirul Qhalis bin Abu Rashid, Mr. Mohd Najib bin Temizi).

Conflict of interest

There is no any conflict of interest to this chapter as far as the authors' knowledge during the submission.

Author details

Haylay Tsegab[1,2*] and Chow Weng Sum[2]

1 Southeast Asia Carbonate Research Laboratory (SEACaRL), Department of Geosciences, Faculty of Science and Information Technology, Universiti Teknologi PETRONAS, Perak, Malaysia

2 Department of Geosciences, Faculty of Science and Information Technology, Universiti Teknologi PETRONAS, Perak, Malaysia

*Address all correspondence to: haylay.tsegab@utp.edu.my

IntechOpen

References

[1] Malaysia CS, Raj JK. Geomorphology. In: Geology of Peninsular Hutchison and D. 2009

[2] Haq BU, Hardenbol J, Vail PR. Chronology of fluctuating sea levels since the Triassic. Science. 1987;**235**(4793):1156-1167

[3] Heckel PH. Sea-level curve for Pennsylvanian eustatic marine transgressive-regressive depositional cycles along midcontinent outcrop belt. North America. 1986;**14**(4):330-334

[4] Gale AS, Hardenbol J, Hathway B, Kennedy WJ, Young JR, Phansalkar V. Global correlation of Cenomanian (Upper Cretaceous) sequences: Evidence for Milankovitch control on sea level. Geology. 2002;**30**(4):291-294

[5] Vail PR. Seismic stratigraphy interpretation using sequence stratigraphy: Part 1: Seismic stratigraphy interpretation procedure. 1987

[6] Hearty PJ, Kindler P, Cheng H, Edwards RJG. A+ 20 m middle Pleistocene sea-level highstand (Bermuda and the Bahamas) due to partial collapse of Antarctic ice. Geology. 1999;**27**(4):375-378

[7] Van Wagoner J et al. An overview of the fundamentals of sequence stratigraphy and key definitions. 1988

[8] Miller KG et al. The phanerozoic record of global sea-level change. Science. 2005;**310**(5752):1293-1298

[9] Haq BU, Schutter SR. A chronology of Paleozoic sea-level changes. Science. 2008;**322**(5898):64-68

[10] Vail PR. Seismic stratigraphy and global changes of sea level, Part 5: Chronostratigraphic significance of seismic reflections, seismic stratigraphy—Applications to hydrocarbon exploration. Memoirs of American Association of Petroleum Geologists. 1977;**26**:99-116

[11] Metcalfe I. Tectonic evolution of the Malay Peninsula. Journal of Asian Earth Sciences. Oct 25 2013;**76**(0):195-213 (in English)

[12] Tucker E. Sedimentary Petrology: An Introduction to the Origin of Sedimentary Rocks. Wiley; 2009

[13] I. Metcalfe, "The Bentong–Raub Suture Zone," Journal of Asian Earth Sciences, vol. 18, no. 6, pp. 691-712, 12 2000

[14] Lee CP. Palaeozoic Stratigraphy. In: Hutchison CS, Tan DNK, editors. Geology of Peninsular Malaysia. 1st ed. Malaysia: Geological Society of Malaysia; 2009. pp. 55-86

[15] I. Metcalfe, "Late palaeozoic and mesozoic palaeogeography of Southeast Asia," Palaeogeography, Palaeoclimatology, Palaeoecology, vol. 87, no. 1-4, pp. 211-221, 10// 1991

[16] Harbury NA, Jones ME, Audley–Charles MG, Metcalfe I, Mohamed KR. Structural evolution of Mesozoic Peninsular Malaysia. Journal of the Geological Society. February 1, 1990;**147**(1):11-26

[17] Foo KY. The paleozoic sedimentary rocks of peninsular Malaysia—Stratigraphy and correlation. In: Workshop on Stratigraphic Correlation of Thailand and Malaysia Thailand; vol. 1. Thailand: Geological Society of Thailand; 1983. pp. 1-19

[18] Lee CP, Mohamed SL, Kamaludin H, Bahari MN, Rashidah K. Stratigraphic Lexicon of Malaysia. 1st ed. Malaysia: Geological Society of Malaysia; 2004 p. 176

[19] Meor HH, Lee CP. The Devonian-lower carboniferous succession in Northwest Peninsular Malaysia. Journal of Asian Earth Sciences. Mar 2005;**24**(6):719-738 (in English)

[20] Jones CR. The geology and mineral resources of the Girk area, Upper Perak. In: Geological Survey West Malaysia District Memoir. 1st ed. New York; 1973

[21] Hutchison CS. Geological Evolution of Southeast Asia. 2nd ed. Kuala Lumpur: Geological Society of Malaysia; 1996

[22] Suntharalingam T. Upper palaeozoic stratigraphy of the area west of Kampar, Perak. Geological Society of Malaysia. February 1968;**1**:1-15

[23] Ingham FT, Bradford EF. The Geology and Mineral Resources of the Kinta Valley. 1st ed. Malaysia: Geological Survey Headquarters; 1960 pp. 1-77

[24] De Morgan JJM. Note sur la géologie et sur l'industrie minière du Royaume de Pérak et des pays voisins. Ch. Dunod; 1886

[25] Wray L. Some accounts of the Tin-mines and Tin-mining industries of Perak; 1894

[26] Scrivenor JB. The Gopeng Beds of Kinta (Federated Malay States). Quarterly Journal of the Geological Society. 1912;**68**(1-4):140-163

[27] Scrivenor JB. The geological history of the Malay Peninsula. Quarterly Journal of the Geological Society. 1913;**69**(1-4):343-371

[28] Cameron WE. Criticizes Dr. W.R. Jones theory of the geology of the Kinta Valley tin-field. Mining Journal. 1924;**144**:171

[29] Cameron W. The deep leads of Kinta Valley. Mining Magazine. 1924;**31**(b):276-285

[30] Haylay TG, Hunter WA, Chow WS. Depositional environment of the Kinta Limestone, Western Peninsular Malaysia. In: AAPG International Conference & Exhibition Istanbul, Turkey; 2014; AAPG. 2014. p. 36

[31] Gebretsadik HT, Sum CW, Yuriy GA, Hunter AW, Ab Talib J, Kassa S. Higher-resolution biostratigraphy for the Kinta Limestone and an implication for continuous sedimentation in the Paleo-Tethys, Western Belt of Peninsular Malaysia. Turkish Journal of Earth Sciences. 2017;**26**(5):377-394

[32] Tucker ME. Techniques in Sedimentology. Oxford England: Blackwell Science; 1988

[33] Bailey SW. Practical notes concerning the indexing of X-ray powder diffraction patterns of clay minerals. Clays and Clay Minerals. 1991;**39**(2):184-190

[34] Dong C. PowderX: Windows-95-based program for powder X-ray diffraction data processing. Journal of Applied Crystallography. 1999;**32**(4):838

[35] Killops SD, Killops VJ. Introduction to Organic Geochemistry. 2nd ed. USA: Wiley; 2013. p. 408

[36] Kennedy MJ, Lohr SC, Fraser SA, Baruch ET. Direct evidence for organic carbon preservation as clay-organic nanocomposites in a Devonian black shale; from deposition to diagenesis. Earth and Planetary Science Letters. 2014;**388**(0):59-70 (in English)

[37] Rey J, Galeotti S. Stratigraphy: Terminology and Practice (no. Editions Technip). Comité français de stratigraphie; 2008

[38] Coe AL. The Sedimentary Record of Sea-Level Change. Open University; 2003

[39] Weissert H, Joachimski M, Sarnthein M. Chemostratigraphy. Newsletters on Stratigraphy. 2008;**42**(3):145-179 (in English)

[40] Blatt H, Tracy R, Owens B. Petrology: Igneous, Sedimentary, and Metamorphic. 2nd ed. New York, USA: W.H. Freeman; 2006

[41] McIlreath IA, James NP. In: Walker RG, editor. Facies Models. Vol. 5. Geological Association of Canada; 1984. pp. 189-199

[42] Alsop GI, Marco S. A large-scale radial pattern of seismogenic slumping towards the Dead Sea Basin. Journal of the Geological Society. 2012;**169**(1): 99-110 (in English)

[43] Principaud M, Mulder T, Gillet H, Borgomano J. Large-scale carbonate submarine Mass-Wasting along the Northwestern slope of the Great Bahama Bank (Bahamas): Morphology, architecture, and mechanisms. Sedimentary Geology;**2014**

[44] Soares JL, Nogueira ACR, Domingos F, Riccomini C. Synsedimentary deformation and the paleoseismic record in Marinoan cap carbonate of the southern Amazon Craton, Brazil. Journal of South American Earth Sciences. Dec 2013;**48**:58-72 (in English)

[45] Rubert Y, Jati M, Loisy C, Cerepi A, Foto G, Muska K. Sedimentology of resedimented carbonates: Facies and geometrical characterisation of an upper Cretaceous calciturbidite system in Albania. Sedimentary Geology. 2012;**257**:63-77 (in English)

[46] Waldron JWF, Gagnon JF. Recognizing soft-sediment structures in deformed rocks of orogens. Journal of Structural Geology. 2011;**33**(3):271-279 (in English)

[47] Alsop GI, Marco S. Soft-sediment deformation within seismogenic slumps of the Dead Sea Basin. Journal of Structural Geology. 2011;**33**(4):433-457 (in English)

[48] Kawakami G, Kawamura M. Sediment flow and deformation (SFD) layers: Evidence for intrastratal flow in laminated muddy sediments of the Triassic Osawa Formation, northeast Japan. Journal of Sedimentary Research. 2002;**72**(1):171-181

[49] Kennedy W, Juignet P. Carbonate banks and slump beds in the Upper Cretaceous (Upper Turonian-Santonian) of Haute Normandie, France. Sedimentology. 1974;**21**(1):1-42

[50] Allen J. Sedimentary Structures, Their Character and Physical Basis. The Netherlands: Elsevier Science; 1982. p. 662

[51] Boggs S. Petrology of Sedimentary Rocks. 2nd ed. Cambridge University Press; 2009. p. 600

[52] Cocks LRM, Fortey RA, Lee CP. A review of lower and middle palaeozoic biostratigraphy in west peninsular Malaysia and southern Thailand in its context within the Sibumasu Terrane. Journal of Asian Earth Sciences. 2005;**24**(6):703-717 (in English)

[53] Metcalfe I. Devonian and carboniferous conodonts from the Kanthan Limestone, Peninsular Malaysia and their stratigraphic and tectonic implications. Memoir-Canadian Society of Petroleum Geologists. 2002;**19**:552-579

[54] Vandeginste V, John CM. Diagenetic implications of stylolitization in pelagic carbonates, Canterbury Basin, Offshore New Zealand. Journal of Sedimentary Research. 2013;**83**(3):226-240

[55] Moore CH. Carbonate Reservoirs: Porosity Evolution and Diagenesis in a Sequence Stratigraphic Framework. Elsevier Science; 2001

[56] Sunagawa I, Takahashi Y, Imai H. Strontium and aragonite-calcite precipitation. Journal of Mineralogical and Petrological Sciences. 2007;**102**(3):174-181 (in English)

[57] Fairbridge RW, Bissel HJ, Chilingar GV. Carbonate Rocks (no. pt. 1). Amsterdam: Elsevier Science; 1967

[58] Hutchison CS. Physical and chemical differentiation of West Malaysian limestone formations. Geological Society of Malaysia. 1968;**1**:45-56

[59] Chilingar GV, Bissell HJ, Fairbridge RW. Carbonate Rocks (Development in Sedimentology A). Amsterdam: Elsevier Science; 1967

[60] Boggs S. Principles of Sedimentology and Stratigraphy. 4th ed. America: Pearson Education Ltd.; 2006

Section 4

Chronostratigraphy

High Resolution Chronostratigraphy from an Ice-Dammed Palaeo-Lake in Andorra: MIS 2 Atlantic and Mediterranean Palaeo-Climate Inferences over the SE Pyrenees

Valenti Turu

Abstract

This work is close beyond the common applicability of the sequence stratigraphy principles. Nevertheless one aim is to demonstrate its validity for a small ice-dammed palaeo-lake. Since base-level variability is smoothly related with an oscillatory motion, the chronostratigraphy chart emerges as a crucial tool to produce enough quantitative data for a Fast Fourier Transform (FFT) analysis, even if this chronostratigraphy is based on local stratigraphic constraints. The exposed case is related to sixth order stratigraphic cycles, even higher, related to high-frequency global events occurring at the end of the Pleistocene. The glacio-lacustrine record was controlled by an intermittent ice-damming from the main glacier located in Andorra. Tributary glacier tongues advance and retreat accordingly to the main glacier, however asynchronously. The system Tract evolution and the related unconformities reflect the glacier motion imprint. The glacier advances produce an unconformable subglacial till, but also a repeated TST-HST sedimentary evolution over it and their related conformable surfaces. The long-term base-level evolution was leaded by axial tilt cycle, however precession did not appear at the FFT analysis, indicating a strong high latitude climatic influence on the palaeo-lake's evolution. The sub-orbital periodicities have been observed in the range of the Heinrich events, but also in the range of the Mediterranean salinity anomalies (periodicity of 3–5 ky).

Keywords: Andorra palaeo-glaciers, ice-dammed palaeo-lake, orbital cycles, Heinrich events, Mediterranean moisture, palaeo-climate

1. Introduction

The used approach allows to use the principles of sequence stratigraphy in order to study the Valira del Nord sedimentary infill, which has been controlled by the activity of the valley glaciers.

The unstable glacier motion is identified from its footprint in the sedimentary record. This non-stability was forced by global events [1]. Thus, the derived sedimentary changes can be then considered as a climate proxy. In such a case, the sedimentary infill is ruled by sixth order stratigraphic cycles [2] or higher frequency global events, like Heinrich events. In such natural phenomena the icebergs break off from the glaciers crossing the Northern Atlantic Ocean influencing the circulation patterns of thermohaline currents.

The sedimentary infill from the La Massana has been assimilated to eustasy cycles produced in a passive margin basin. In such a case the deltaic build-up depends exclusively on base level changes. In the case of La Massana, an ice-dammed palaeo-lake was produced at each significant glacier advance.

The mapping resolution derives from available profiles (about 232 profiles) par basin surface (about 11 km^2), here close to 50 profiles par km^2. Stratigraphic data is accessible in an online database [3]. The chronological resolution derives from previous published dating [1], from about 17 absolute dating in a span of time of 46 cal ka (one par 3 ka).

Previous woks [4] supposed a regular timing of the depositional sequence build-up, however this assumption was supported with too few datings. Although the impact of global events in continental areas, like Heinrich events, was still not well known [5]. However increasing data, especially geochemistry and absolute datings, a first correlation to the global events could be performed by [6, 7]. Nowadays, since the chronostratigraphic framework has been settled at La Massana [1] it is possible to assess numerically the impact of the global events.

This work aims to explore about the main forcing causes that ruled the system tracts deposition in this ice-dammed palaeo-lake.

2. Setting

The Principality of Andorra (42°30′N, 1°30′E) is located in the southern slope of the eastern Pyrenees (**Figure 1**). The height of about one half of its surface (233 km^2) exceeds 2000 m a.s.l., and the landscape has been mainly controlled by the glacial geomorphology and by glacial geology. The tributary valley system has a "Y" shape, with two main basins joining in the city of Andorra (1000 m a.s.l.). The highest peaks occur in the Valira del Nord valley, the NW tributary (145 km^2), reaching 2942 m a.s.l. (Coma pedrosa Peak), not far from the first easternmost 3000 peak from the Pyrenees, the Pica d'Estats peak (3143 m a.s.l.) on the Franco-Spanish border. In former times different glaciers coming from the Valira del Nord valley converged with those coming from the NE valleys (211 km^2), and form the main glacier of Andorra (**Figure 2**). The NE tributary embraces the Valira d'Orient (166 km^2) and the Madriu valley (45 km^2), the last one listed in the UNESCO's world human heritage. This main glacier overflowed the Spain-Andorra border (838 m a.s.l.) at the Upper Pleistocene [8].

The studied area is located in the lowest part of the Valira del Nord tributary (**Figure 3**), involving the village of Erts (42° 33′ 41″ N–1° 29′ 48″ E, 1360 m a.s.l.), and the towns of La Massana (42° 32′ 44″ N–1° 30′ 53″ E, 1230 m a.s.l.) and Ordino (42° 33′ 19″ N–1° 31′ 60″ E, 1300 m a.s.l.). The Valira del Nord population (15,000 people) lives in 1% of its basin surface and urban zones lack in space. Understanding the stratigraphy and the extent of the palaeo-lake sediments is quite relevant for Engineering-Geologists who work in La Massana. Glaciolacustrine deposits produced during the Last Glacial Period influence mainly on building bearing capacity and hazards of excavations. For this reason when the spatial distribution knowledge is improved, the variability and predictability nature of sediments is reduced. Geological knowledge improves the safety factors to apply on such projects and reduces costs. No-negligible financial efforts are actually employed

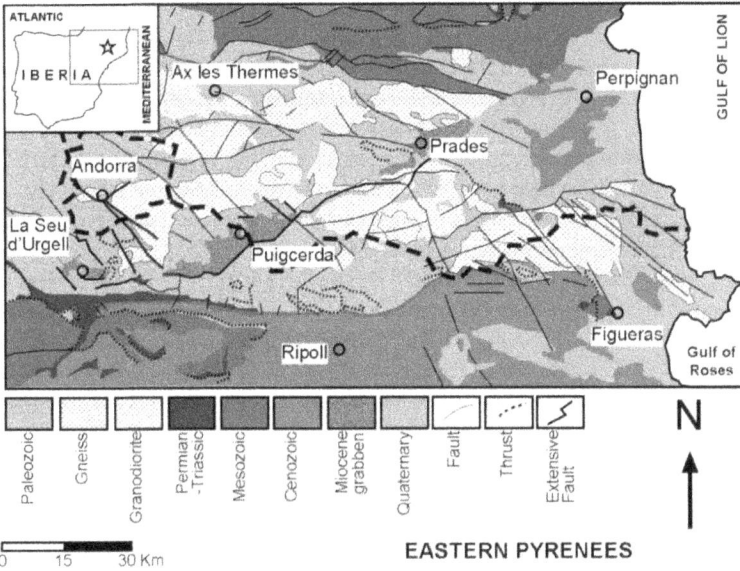

Figure 1.
Geological framework from the eastern Pyrenees, rich in metamorphic and granite rocks from the Paleozoic. Mesozoic rocks are mainly located side to side from the Paleozoic mountain range, however Cenozoic rocks are discordantly placed over the Paleozoic rocks close to Prades, Puigcerda and La Seu d'Urgell. Such rocks are related to a large graben system until the Gulf of lion. Andorra is situated on the western side of the eastern Pyrenees.

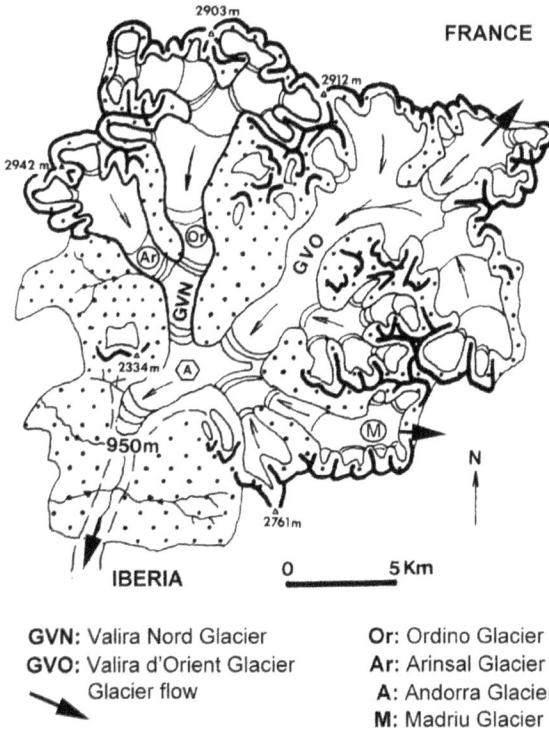

GVN: Valira Nord Glacier   **Or:** Ordino Glacier
GVO: Valira d'Orient Glacier   **Ar:** Arinsal Glacier
        Glacier flow   **A:** Andorra Glacier
        M: Madriu Glacier

Figure 2.
General palaeo-glaciers configuration from Andorra. At the upper Pleistocene three large valleys converge at the same point to form the main glacier of Andorra.

for geotechnical surveying, what benefit the sequence stratigraphy approach by the available profiles. Fortunately or unfortunately, it depends on the lecturer's opinion, there is no any other region in the southern slope of the Pyrenees with a comparable building pressure than in Andorra, making La Massana palaeo-lake a case unique in this region.

However, unique are also other ice-damming cases in the Pyrenees, like Llestui, Llauset, Cerler, Linás de Broto [9, 10], or in their respective mountain massif, like Pias in NW Spain [11], but in none of such cases had been correlated to global climatic events previously as in La Massana [1].

La Massana palaeo-lake was dammed by the largest glacier from Andorra at MIS 3 [12], long time after the maximum ice extent (MIE) in Andorra (**Figure 4**). The MIE is previous to 59 ka, just before a widespread glacier recession that starts before 40 ka cal BP [13]. The maximum ice extent (MIE) in the Pyrenees is diachronous to the global last glacial maximum (LGM) of the northern hemisphere ice sheets [14–18].

This large period of glacier recession in the Pyrenees at the MIS 3 start is a classical boundary from the local würmian stratigraphy of prehistoric excavations, the würm II - würm III boundary [19, 20]. A general glacier advance before 32 ka occur in Andorra [21], also in the SW Pyrenees [22], and in the inner Iberia could be the beginning of glacier tongues presence [23]. The starting point of the glacio-lacustrine sedimentation in La Massana palaeo-lake is related to that general glacier advance at MIS 3. However all glaciers recede again short after [1]. Only the SE Pyrenees glaciers recover their original extent at MIS 2 [1], but those from the SW Pyrenees not [24].

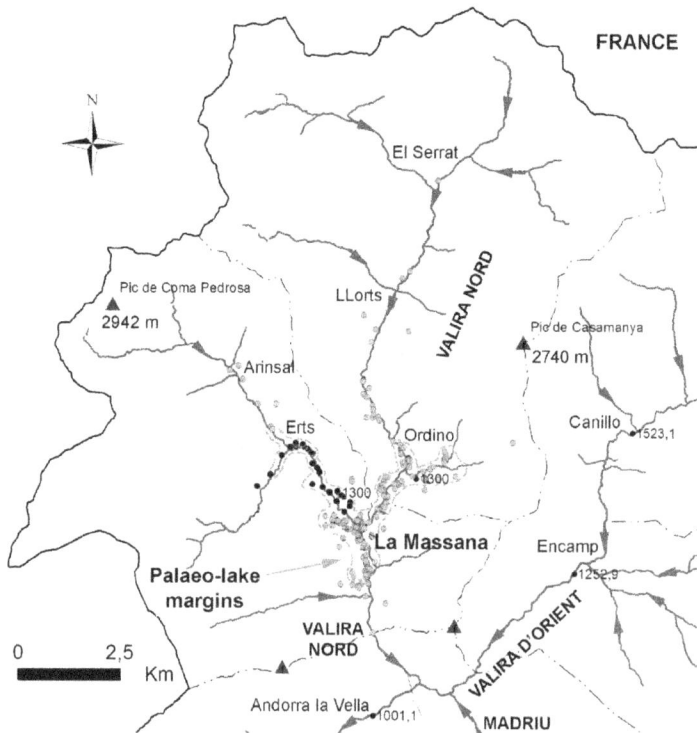

Figure 3.
Extension of the palaeo-lake and related stratigraphic profiles. The profiles related to the Erts and Hortals deltas are plotted as black dots.

LA MASSANA PALEOLAKE

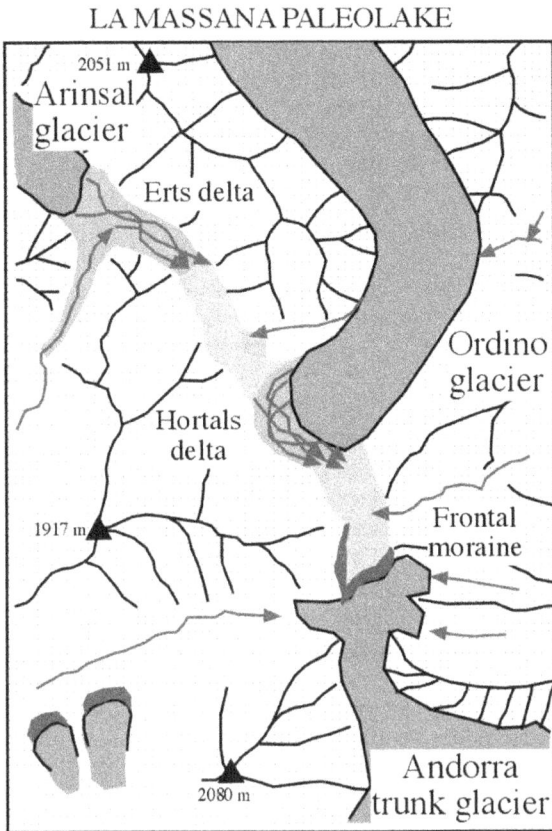

Figure 4.
La Massana palaeo-lake configuration sketch pad. The Valira del Nord valley get trunk by the main glacier of Andorra. This produced the formation of the La Massana palaeo-lake. Local glaciers (Ordino and Arinsal glaciers) build two proglacial deltas, the Hortals and Erts deltas.

3. Methods

Two different but complementary disciplines of the sequential stratigraphy can be clearly differentiated: 1) the analytical way and 2) the synthetic way. The last of both aims to develop a time scale of global changes that can be inserted into the global chronostratigraphic scale. The first one, that is the analytical concept, is understood as stratigraphic modeling of facies associations.

In any cases the main goal is the recognition of units (set of strata) within the basin infill materials, limited by surfaces that mark a genetic change conditions affecting the entire basin. Its recognition is a mandatory object in all basin analysis.

3.1 Hierarchy

The applicability of the sequence stratigraphy to large ice-dammed palaeo-lakes has been settled since the end of the last century [25]. Shortly after sequence stratigraphy principles were used in Andorra [21]. In such works the sequence stratigraphy terminology is widely applied such as depositional sequences, system tracts and parasequences. However, since theses pioneer works the use of the sequence stratigraphy terminology to describe the sedimentary architecture from glaciated valleys in Iberia starts to be common [1, 8, 17, 26].

The term of depositional sequence were introduced by [27, 28], and corresponds to a relatively concordant stratigraphic succession of genetically related strata, where the base and the top are discontinuities, as well as their correlative surfaces though the basin. However, the essential criteria for the recognition and subdivision of sequences were established by [29]. Two kind of sequence limits are established by discontinuities (SB 1 and SB 2, Surface Boundary), what define also the depositional sequence type.

The System Tracts were introduced by [27, 28] to name the set of contemporaneous depositional systems formed under the same base level conditions. The System Tracts sedimentary models were established according to their transgressive retrograde character (TST, transgressive system tract), regressive prograde (FSST, falling stage system tract & LST, lowstand system tract), transgressive or prograde character (HST, highstand system tract).

A set of relatively concordant strata limited by flooding surfaces are parasequences [30]. Three types of parasequences can be distinguished: (1) Those that had a progradational character, (2) those that are retrogradational, and (3) those that had an aggradational character. Such sedimentary build-up is according to the material supply contribution, but also in function to the base level evolution.

Since this approach is based on the magnitude of base-level changes and related boundary formation (independently to the cycle span), sequence ordering are somewhat relative. Ordering is based on the principle that a sequence cannot involve a boundary that has an equal or greater magnitude than the one at the start of the sequence [31], but also the magnitude of base-level changes leads the hierarchical rank of the bounding surface (SB, sequence boundary) and unconformities (US, unconformity surface), so temporal and spatial scales are not relevant [32], and thus applicable to this case of study.

3.2 Dichotomization

Since glacial valleys form by external processes, the sedimentary infill at their troughs, can be assimilated to the one produced in a passive margin [4]. Thus, only type 1 discontinuity can able to be produced in constrained glacial valleys like La Massana. Sequences starting on such kind of discontinuity had three system tracts, a FSST-LST, the TST and HST. However for large ice-dammed palaeo-lakes type 2 discontinuity has been announced by [25].

In the case of constrained ice-dammed palaeo-lakes from valley glacier, the glacial tongues were small comparing to the mountain range rot (>25 km) [33], what

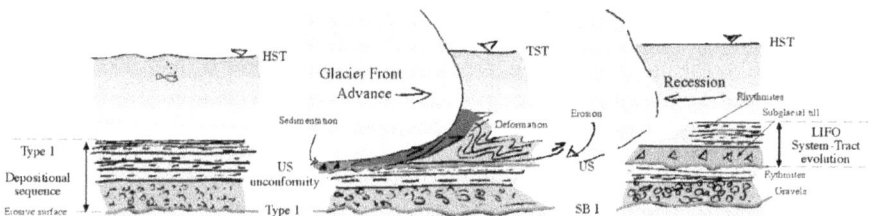

Figure 5.
Subglacial till unconformity and LIFO (last in-first out) system tract evolution. From left to right: The trunk glacier flood the palaeo-lake produces a type 1 depositional sequence with almost three system tracts (LST, TST, HST). The local glacier invaded the palaeo-lake, deformed and eroded previous sediments; a subglacial till is deposited; and the overall produced an unconformity. Recession of the local glacier promoted the formation of two new system tracts (TST and HST), over the unconformity surface and its correlative surfaces. In such evolution the last glacier that came in La Massana's palaeo-lake it was also the first to left (LIFO) the palaeo-lake.

inhibit quick glacio-isostatic rebound and vertical accommodation, reducing the possible surface boundaries to an unique case, the type 1 case. For large ice-dammed lakes the principles are the same, however several sedimentary assemblages types can coexist [25], in parallel to sedimentary accommodation [34] or uplift (glacio-isostatic rebound).

A particular unconformity has been defined [4] related to the glacier motion. When the glacier tongue overrode its own proglacial deposits, deformation-compaction and eventually erosion would be produced [25]. In such a case unconformity and diamicton are synchronously produced (**Figure 5**), and thus sequences that include diamictons had two supplementary system tracts, the TST and the HST [1].

3.3 Particular assemblages

Glaciolacustrine deltas in ice-dammed valley glaciers [1, 26] are commonly of Gilbert type [35]. Their bottomset is much larger than the foreset and topset, what mean that most of the available outcrops are related with dense turbidites and rhythmites (**Figure 6**). For that reason the study of fine grain-size facies and their relative position are important. Ice-dammed lakes sedimentation is dominated by turbidity currents when the ice dam seals the end of the valley [36, 37].

If the damming increases an induced transgression occurred by ice thickening, then lacustrine facies migrated upward into the valley. Glacier induced

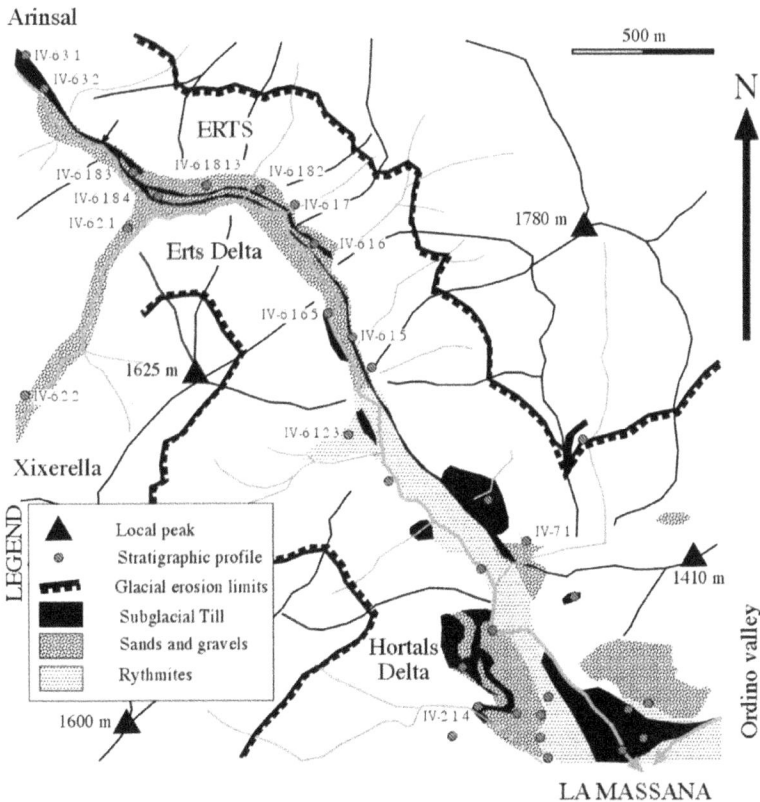

Figure 6.
Simplified sedimentary mapping facies from the La Massana palaeo-lake between the Hortals and Erts deltas and related stratigraphic profiles [3]. The Hortals delta overrode the bottom set of the Erts delta. Subglacial till layers interdigite delatic sediments.

unconformity drive base level to shift, similarly as by an eustacy cause [29]. Once the ice ends to grow, the base level remains stable and delta progression is due by large outwash from meltwaters [17, 26]. Mass flows and grain flows on the foreset slope are then produced [36]. Inversely graded gravel and sand facies were found by [4] filling scours from previous mass flow channels, similar to the high stand turbidites [38], but also lenticular bodies that had Bouma's truncated series [39] were described at the bottomset [1].

If the damming ceased by glacier recession, the base lake level drops producing very marked erosion and incisions [32]. Debris flows and denser turbidites are then deposited in the lowstand fans [38].

4. Sedimentary architecture from La Massana palaeo-lake

Two deltas have been distinguished (**Figure 6**), the Hortals and Erts deltas, forming an staircase system that works sequentially. The sedimentary build-up of the Hortals delta trunk the bottom set of its tributary's delta (Erts), however truncation was unstable by the glaciers front oscillations. When the local glacier

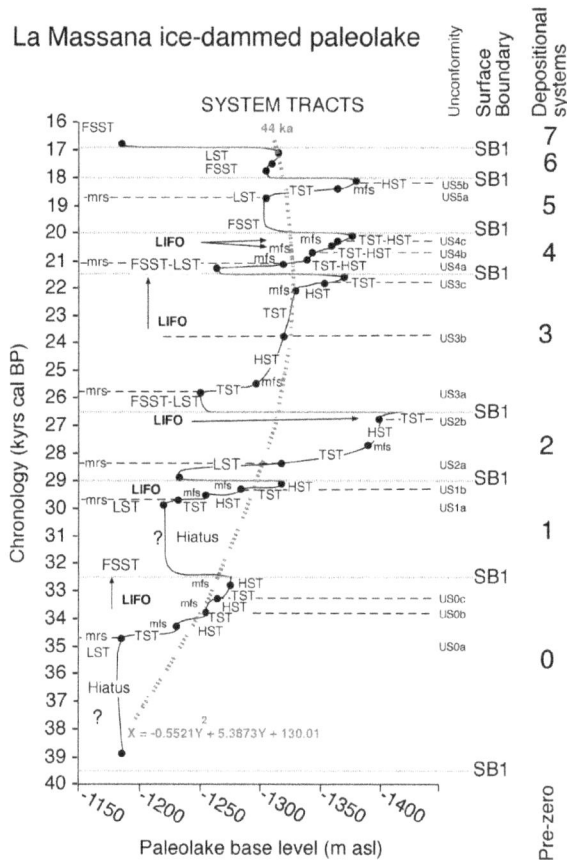

Figure 7.
Simplified chronostratigraphy from the La Massana palaeo-lake inferred from the base-level evolution. SB: Sequence boundary. US: Unconformity surface. FSST, forced stand system tract; LST, low system tract; TST, transgressive system tract; HST, high system tract, LIFO, last in-first out system tract evolution; mrs, maximum regressive surface; mfs, maximum flooding surface. Overprinted the polynomial evolution of the overall base-level evolution (parabola).

stabilized its front at La Massana palaeo-lake, the Hortals delta gets feed with outwash sediments coming from the glacier's meltwaters. When the Ordino glacier advanced over the Hortals delta its tributary (Arinsal valley) was then flooded and a highstand proglacial delta was built at the locality of Erts, feed by meltwaters of Arinsal glacier [1]. Glacier recession is produced in the reverse order, the last to come was the first to quit the palaeo-lake's margin.

Almost seven depositional sequences (SD) have been described in the valley of Arinsal [1], each divided from the others by type 1 discontinuities (**Figure 7**). In all cases, the nature of the discontinuity leads by subaerial exposure and colluvium deposits (FSST). In some of the recessional phases, the LST local base level could lead by the main glacier of Andorra, the Valira d'Orient glacier. Then coarse turbidites outward from the prodelta feeding with sediments the bottom set (SD2 and SD4) [1].

In all depositional sequences the TST starts when the ice volume in the Valira d'Orient trunk glacier increase and flood its tributary valley. As damming persisted, parasequences became retrograde and low-density turbidites where produced. Ice-damming steady-state and maximum flooding surfaces (msf) are both linked, regarding the start of the HST. Extensive rhythmites are then deposited in the palaeo-lake. However if damming progresses a higher base-level reach in the palaeo-lake, subglacial till sedimentation also progresses by the advance of the local glaciers. This subglacial till is a sedimentary unconformity (Unconformity Surface, US) correlative to a transgressive surface (TS), or maximum regressive surface (mrs), promoting the start of a new TST [1].

The two starting sequences (**Figure 7**) had relative large sedimentary hiatus starting abruptly in a Forced Stand System Tract (FSST). That happen also at the start of the third and fifth sequences (**Figure 7**), however the time span is lesser at the start of the second and the forth sequence (**Figure 7**).

The relationship between water-depth changes is a complex matter, and not always elucidated [32]. Being aware to this, only absolute height of the base-level (meters above the sea level) is here considered. When applying this to the general trend of the palaeo-lake base-level for all HST and all LST (**Figure 8**) a normal regression evolution [32] is identified at both. From this is possible to infer that deltas topsets grow faster than their bottomsets [32], or differential flooding is forded

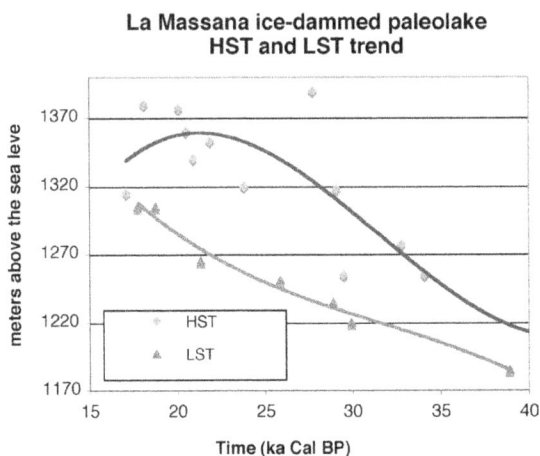

Figure 8.
HST and LST base-level evolution from the La Massana palaeo-lake. In absence of subsidence this evolution indicates shallowing along the entire profile [32]. The maximum flooding surfaces forms in deepening waters [32]. The surfaces corresponding to the peak of deepest water forms within the LIFO system tract evolution.

by the glaciers (local ice-tongues melt faster). Either cases promotes the formation of long clinoforms through time [40, 41], resulting in thin bottomsets and thick delta fronts (foresets and topsets).

Several TST and HST had been distinguished [1] within the sequences. Three intra-sequence unconformities (US) are identified regarding the glaciers motion. The trunk glacier (Andorra) produces the first unconformity (USa), and its correlative surface is a maximum regressive surface (mrs). The biggest local glacier (Ordino) produces a second unconformity (USb) when it advance, and its correlative surface is a transgressive surface (TS). The last unconformity (USc) is produced when the Arinsal (the smallest glacier) advance into the palaeo-lake. However at the younger sequences (SD7 and SD6) the Arinsal glacier did not reach the palaeo-lake, only USa and USb were formed [1].

5. Depicting base-level rates

Low-order polynomial curves had been used to fit long-term trends in time series data, like first-order functions (linear) and second-order polynomial quadratic type function:

$$y = ax^2 + bx + c \qquad (1)$$

Wich is a parabola involving a single maximum value. Progressive increase in complexity, nth-order polynomial equation had n-1 maxima and minima values, that can be used to extract trends from a log evolution (Figure base level), like the palaeo-lake base-level. The rate of the base-level change for a given time-step is given by the first derivative of the polynomial Equation [32], while the second derivative the variation of the rate for a given time-step (acceleration).

For La Massana glaciolacustrine record two maxima are almost needed to ensure the general trend for the palaeo-lake base-level, one at the MIS 2 and the other one at the MIS 4 related to the MIE glacial phase [1]. The minimal base-level happen when the general glacier recession occur at the beginning of the MIS 3 [1]. Then a

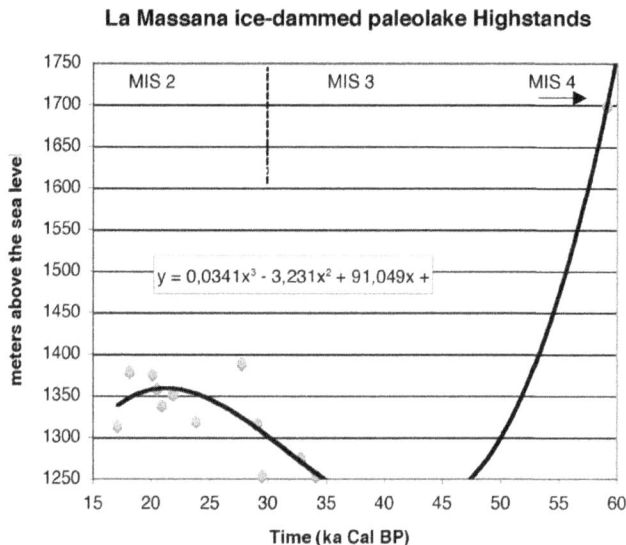

Figure 9.
Polynomial adjustment for the evolution of the palaeo-lake's HST between MIS 3 and MIS 2.

First and Second derivative HST function

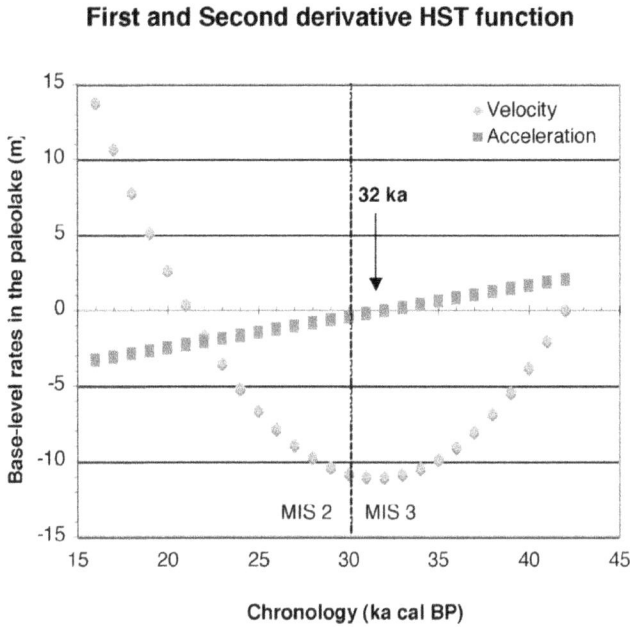

Figure 10.
Derivative rates from the polynomial equation involving all the palaeo-lake's HST. The first derivative (velocity) reacts to the negative trend of the second derivative (deceleration). Motion's rates change at the MIS 3- MIS 2 boundary. Less frequent D-O events promote persistent LIFO system tract evolution and higher base-levels through time (see text for further explanation).

third-order polynomial equation (n-1 = 2) must be invoked first (**Figure 9**). The first derivative (base-level rate) is a parabola (**Figure 10**), so a single maximum (a minimum in this case) is expected. The second derivative (variability of the rate) is here a linear function (**Figure 10**) and the acceleration is proportional to its inclination.

A Massana the sense of the acceleration change at 32 ka to a negative decrease, in fact acceleration increases. Thus the base-level rate (velocity) starts to increase, close to the MIS 3 – MIS 2 boundary.

6. Depicting cycles from the sedimentary record

The chronological differences between sequences are related to a high frequency cycles [42] in the palaeo-lake. Time-based hierarchy emphasizes here general orbital mechanisms driving the existing glacier to dam the valley, however the damming intermittency is ranging suborbital frequencies [1].

Fast Fourier Transform spectral analysis from the base-level rate (**Figure 9**) exhibit significant correlation to the orbital 41 ka axial tilting period (**Figure 11**), which affect the total annual solar radiation at high latitude (Imbrie & Imbrie, 1986). However, Andorra's latitude is not so high, but the main trend of the ice-damming acts as it was. Equinoxes precession are not present in the FFT spectra, a 23 ka time period orbital cycle [43], affecting the total annual solar radiation at low latitude [44]. This would imply a strong climate forcing from higher latitude climate over Andorra.

Faourier Transform spectral analysis from the base-level itself (**Figure 7**) exhibit significant correlation to sub orbital periods (**Figure 12**). The 7.4 kyrs period would

Paleolake base-level rate Fast Fourier Transform

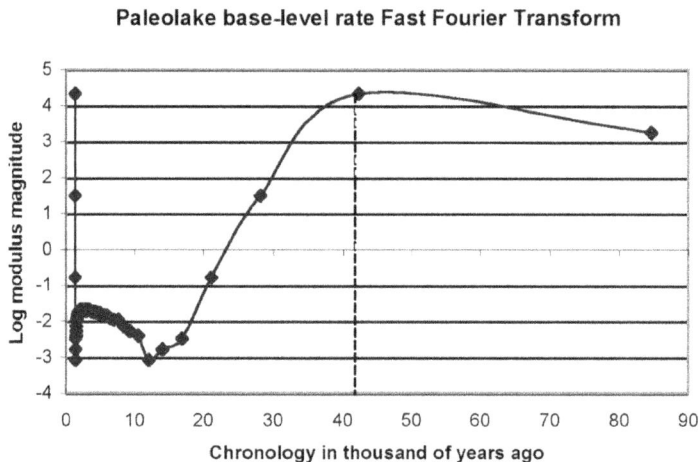

Figure 11.
Fast Fourier transform from the polynomial function (parabola) from the average base-level evolution. Orbital axial tilt frequency is the main Milankovich cycle ruling the palaeo-lake's base-level, and thus ruling the glacier's motion.

Paleolake base-level Fast Fourier Transform

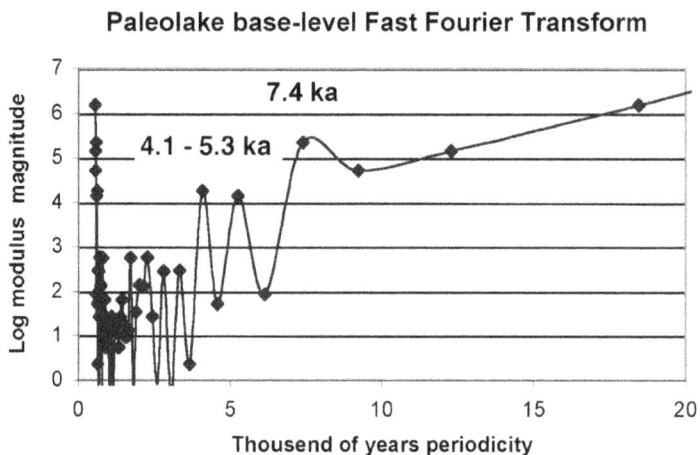

Figure 12.
Fast Fourier transform from the base-level evolution. High frequency cycles arise in the FFT spectrum at 7.4, 5.3 and 4.1 ka. Heinrich events imprint were identified in the palaeo-lake's record [1], and might correspond to the highest peak in the spectrum (7.4 ka). Sub-Milankovich cycles with higher frequencies than the Heinrich events could be related to Atlantic – Mediterranean palaeo-climate inferences (see text for further explanation).

be related to Heinrich events [1], which affect the North Atlantics polar front position. Higher frequency cycles with a period of 4.1 kyrs and 5.3 kyrs are also depicted from the FFT base-level data, linked to strong dry westerly winds blowing over the Gulf of Lion and the strength variability of the western Mediterranean stormy tracks [45].

Theiterranean can produce regional moisture increases by itself, most likely focused on winter. WMDW (Western Mediterranean Deep Water) formation is switched-on by outbreaks of the cold and relatively dry Mistral over the Gulf of Lions. Then Bernoulli aspiration of deep waters outflows through the Strait of Gibraltar [46]. WMDW is produced in winter, however it may not occur every year [45]. Nevertheless

WMDW formation under extreme climate forcing outside the Gulf of Lion [45] could explain the paleoclimatology of the Eastern Pyrenees.

7. Discussion

Since a strong climate forcing from high latitudes rules the Andorra (SE Pyrenees) palaeo-climate, a feedback response may exist within the Mediterranean sea. This high latitude influence must be related with the production of strong westerly winds [46]. Westeries blowing over the Gulf of Lion enhance cyclonic circulation, inducing the WMDW to form in the Balearic basin. Within the gyre centre, mixing convective plumes reach great depths (−2000 m) where the net vertical transport is zero, acting more likely mixing elements [47]. Divergent Ekman upwelling promotes pycnocline shallowing between the surface and intermediate waters [45]. A violent mixing phase starts with an intense evaporation and cooling diminishing the density gradient between intermediate waters in the western Mediterranean (WM), and these from the surface [47]. Quickly the mixed waters sink and spread horizontally at great depth into the Balearic basin what form the WMDW [47]. Once the WMDW reach the Alboran Sea, the WMDW is a westward current along the Moroccan coast promoting evaporation and cooling [48]. Mediterranean Outflow through the Strait of Gibraltar involves dense deep outflow from the Strait of Sicily coming from the eastern Mediterranean (TDW, Tyrrhenian Deep Water) [49].

Progressive surface buoyancy increase reduces new deep-water formation, what might cause reduction of Bernoulli aspiration and potential Mediterranean deep-water stagnation [50]. WMDW properties change only due to diffusion (low Bernoulli aspiration) producing salinity anomalies in WM [51] when WMDW salinity is higher than in the MIW (Mediterranean Intermediate Water). Between the middle Holocene and the LGM the modeled salinity anomalies appear in a periodicity of 3–5 kyrs [50]. Such periodicities are in great accordance to the high frequency cycles observed in the La Massana's palaeo-lake FFT spectra.

However, variability to such general WM pattern increases by massive external water incomings, such as deglaciation phases or levantine monsoon influences. In the first case cold Alpine melt-water flow to the Gulf of Lion enhancing salinity differences causing remarkable changes in the Bernoulli aspiration [50]. Levantine monsoon flooding reduces significant differences in salinity and WMDW strength, organic layers deposition in the WM area are then favored when deep-water injections are inhibited [50]. Both cause Mediterranean sea-level increases and in turn a great Gibraltar's strait opening. This effect favor salinity contrasts reduction and low exchanging velocities by continuous freshening of Atlantic inflow [52].

8. Conclusions

Larger valley would produce larger glacier as its de case for the Andorra glacier [1], but this cannot explain a faster ice growing at the smaller valleys. A sequential westward advance of the glaciers is observed [1] as follows: the Valira d'Orient glacier firstly advance and trunk La Massana, then the Ordino's glacier advance and trunk the Arinsal valley, and at the end the Arinsal glacier advance. Reverse glacier motion also occurred, in an eastward glacier recession. Glacier ongoing-outgoing motion strongly influenced the glaciolacustrine record. Last In-First Out glaciers motion produce a base-level pile-up sequence (a LIFO sequence), in where TST-HST repeat several times in a single depositional sequence.

An eastward valley trend of snow owing to easterly moisture incomes of the Mediterranean sea may explain such westward motion of glaciers. Persistence of snow deflation experienced on the windward slopes [17] would explain a faster ice-grow of the eastern glacier of Andorra. This could be the reason why the Andorra glacier block out always its tributary valley first, even several millennia before than the local glaciers advance until the vicinity of the palaeo-lake. Moisture incomings could have a strong Mediterranean influence [53] at the beginning of each sequence, drifting to the west and producing the observed intra-sequence unconformities evolution (LIFO sequence).

The starting time-period of such dynamics is around 32 ka, when negative acceleration and base-level rate increase (**Figure 10**). This coincides when D-O events [54] become less frequent [8, 17]. An increase of the base-level rate entails also better snow accumulation, just in a steady decline of insolation [8]. Long-term evolution of the palaeo-lake base-level is leaded by the 41 kyr orbital cycle.

Precession did not appear at the FFT spectra, somewhat surprising since Andorra's latitude is not so high. This might indicate a strong high latitude climate influence in the palaeo-lake's evolution.

Suborbital periodicities are also observed in the palaeo-lake base-level FFT spectra, in the range of Heinrich events [1], but also in the range of Mediterranean salinity anomalies (3–5 kyrs). Such anomalies are linked to the Mediterranean Levantine monsoon/flooding that reduces the Western Mediterranean Deep Water strength (WMDW). This inhibits the Tyrrhenian Deep Water (TDW) to advance toward the western Mediterranean over the Strait of Sicily, and the WMDW are not able to reach the Alboran sea. However only during persistent strong westerly winds situation blowing on the Gulf of Lion enhance the WMDW current, inverting the stagnation situation of the western Mediterranean.

Dry mistral winds coming from the Atlantics crossing over Iberia to the Mediterranean force the general retreat of the glaciers from the Pyrenees, almost those from the southern slope. Type 1 surface boundaries are then produced in the La Massana palaeo-lake. Ice-damming and system tracts progression could happen only when persistent moisture incomings feed Andorra, apparently in opposition to a low WMDW. However any Mediterranean moisture influence over the SE Pyrenees does not progress with the same intensity toward the Western Pyrenees [53].

Author details

Valenti Turu
Marcel Chevalier Earth Sciences Foundation Edifici Cultural La Llacuna AD500 – Andorra la Vella Principality of Andorra

*Address all correspondence to: igeofundacio@andorra.ad

References

[1] Turu V, Calvet M, Bordonau J, Gunnell Y, Delmas M, Vilaplana J M, Jalut G. Did Pyrenean glaciers dance to the beat of global climatic events? Evidence from the Würmian sequence stratigraphy of an ice-dammed palaeolake depocentre in Andorra. In Hughes P D, Woodward J C, editors. Quaternary Glaciation in the Mediterranean Mountains. Geological Society, London, Special Publications. 2017;**433**(1);111-136. DOI: 10.1144/SP433.6

[2] Vail PR, Audernard F, Bowman SA, Eisner PN, Perez-Cruz C. The stratigraphic signatures of tectonics, eustasy and sedimentologu-an overview. In: Einsele G, Ricken W, Seilacher A, editors. Cycles and Events in Stratigraphy. Berlin: Springer-Verlag; 1991. pp. 617-659

[3] Marcel Chevalier Earth Science Foundation database [Internet] 2013. Available from http://llacglacial.no-ip.org/. [Accessed: 2018-07-31]

[4] Turu V. Análisi secuencial del delta de Erts. Estratigrafía de un valle glacial obturado intermitentemente. Relación con el último ciclo glaciar. Valle de Arinsal, Pirineos Orientales. Parte I: El método utilizado. Parte II: Aplicación. In: Serrano E, editor. Estudios recientes (2000-2002) en Geomorfología: Patrimonio, montaña y dinámica territorial. Valladolid: Sociedad Española de Geomorfología (SEG)-Departamento de Geografía, Universidad de Valladolid; 2002. pp. 555-574

[5] Grousset F. Les changements abrupts du climat depuis 60000 ans. Quaternaire. 2001;**12**(4):203-211. DOI: 10.3406/quate.2001.1693

[6] Turu V, Bordonau J. Estudio geoquímico de los sedimentos glaciolacustres de la Massana y Ordino (Andorra, Pirineos Orientales): Influjo sedimentario entre lagos de obturación yuxtaglaciar, interpretación paleoambiental. In: Baena R, Fernández JJ, Guerrero I, editors. El Cuaternario Ibérico: Investigación en el siglo. Vol. XXI. Sevilla: Asociación Espanola para el Estudio del Cuaternario & Grupo do Trebalho Portugués para ou Estudio do Quaternário (AEQUA-GTPEQ); 2013. pp. 204-208

[7] Turu V, Jalut G. Glaciation in the eastern Pyrenees between NGRIP/GRIP GS-9 and GS-2a stadials: 25,000 years of palaeoenvironmental record in the Valira valleys (Principality of Andorra). In: Proceedings of the International Conference Paléoclimats et environnements quaternaires, quoi de neuf sous le soleil ? (Q10); 16-18 February 2016. Bordeaux: AFEQ-CNF-INQUA; 2016

[8] Turu V, Boulton GS, Ros X, Peña-Monné JM, Bordonau J, Martí-Bono C, et al. Structure des grands bassins glaciaires dans le nord de la péninsule ibérique: Comparaison entre les vallées d'Andorre (Pyrénées orientales), du Gallego (Pyrénées centrales) et du Trueba (Chaîne Cantabrique). Quaternaire. 2007;**18**(4):309-325. DOI: 10.4000/quaternaire.1167

[9] Bordonau J. Els complexos glacio-lacustres relacionats amb el darrer cicle glacial als Pirineus. Geoforma Ediciones: Logroño; 1992. 251 p

[10] Sancho C, Arenas C, Pardo G, Peña-Monné JL, Rhodes EJ, Bartolomé M, et al. Glaciolacustrine deposits formed in an ice-dammed tributary valley in the south-Central Pyrenees: New evidence for late Pleistocene climate. Sedimentary Geology. 2018;**366**:47-66. DOI: 10.1016/j.sedgeo.2018.01.008

[11] Pérez-Alberti A, Valcárcel-Díaz M, Martini IP, Pascucci V, Andreucci S. Upper Pleistocene glacial valley-junction sediments at Pias, Trevinca Mountains, NW Spain. In: Martini IP, French HM, Pérez-Alberti A, editors. Ice-Marginal and Periglacial Processes and Sediments.

Vol. 354. Geological Society, London, Special Publications; 2011. pp. 93-110. DOI: 10.1144/SP354.6

[12] van Meerbeeck CJVC, Renssen H, Roche DMVAP. How did marine isotope stage 3 and last glacial maximum climates differ? 2009 perspectives from equilibrium simulations. Climate of the Past Discussions. 2009;**4**(5):33-51. DOI: 10.5194/cp-5-33-2009, 2009

[13] Planas X, Corominas J, Vilaplana JM, Altimir J, Torrebadella J, Amigó J. Noves aportacions al coneixement del gran moviment del Forn de Canillo, Principat d'Andorra. In: Turu V, Constante A, editors. El Cuaternario en España y áreas afines, avances en 2011. Andorra la Vella: Asociación Española para el Estudio del Cuaternario & Fundació Marcel Chevalier (AEQUA-FMC); 2011. pp. 163-167

[14] Jalut G, Delibrias G, Dagnac J, Mardones M, Bouhours M. A palaeoecological approach to the last 21 000 years in the pyrenees: The peat bog of Freychinede (alt. 1350 m, Ariege, South France). Palaeogeography, Palaeoclimatology, Palaeoecology. 1982;**40**(4):321-336 and 343-359. DOI: 10.1016/0031-0182(82)90033-5

[15] Jiménez-Sánchez M, Farias-Arquer P. New radiometric and geomorphologic evidences of a last glacial maximum older than 18 ka in SW European mountains: The example of Redes Natural Park (Cantabrian Mountains, NW Spain). Geodinamica Acta. 2002;**15**(1):93-101. DOI: 10.1016/S0985-3111(01)01081-6

[16] García-Ruiz JM, Valero-Garcés BL, Martí-Bono C, González-Sampériz P. Asynchroneity of maximum glacier advances in the central Spanish Pyrenees. Journal of Quaternary Science. 2003;**18**:61-72. DOI: 10.1002/jqs.715

[17] Jalut G, Turu V, Dedoubat JJ, Otto T, Ezquerra J, Fontugne M, et al. Palaeoenvironmental studies in NW Iberia (Cantabrian range): Vegetation history and synthetic approach of the last deglaciation phases in the western Mediterranean. Palaeogeography, Palaeoclimatology, Palaeoecology. 2010;**297**:330-350. DOI: 10.1016/j.palaeo.2010.08.012

[18] Hughes PD, Gibbard PL, Ehlers J. Timing of glaciation during the last glacial cycle: Evaluating the concept of a global 'last glacial maximum'(LGM). Earth-Science Reviews. 2013;**125**:171-198. DOI: 10.1016/j.earscirev.2013.07.003

[19] de Lumley-Woodyear H. Le Palèolithique infèrieur et moyen du Midi méditerranéen dans son cadre géologique. Gallia Préhistoire. 1969. V(I-II); 234-445. DOI: 10.1017/S0079497X00012871

[20] Renault-Miskovsky J. Relation entre les spectres archéo-polliniques du Sud-Est de la France et les oscillations climatiques entre 125000 ans et le maximim glaciaire. Quaternaire. 1986;**23**(1-2):56-62. DOI: 10.3406/quate.1986.1793

[21] Jalut G, Turu V. La végétation des Pyrénées françaises lors du dernier épisode glaciaire et durant la transition Glaciaire-Interglaciaire (Last Termination). Els Pirineus i les àrees circumdants durant el Tardiglaciar. In: Proceedings of the XIV Col. loqui internacional d'Arqueologia de Puigcerdà: Mutacions i filiacions tecnoculturals, evolució paleoambiental (16 000-10 000 BP). Puigcerdà: Institut d'Estudis Ceretans; 2008. pp. 129-150

[22] Lewis CJ, McDonald EV, Sancho C, Peña-Monné JL, Rhodes EJ. Climatic implications of correlated upper Pleistocene and fluvial deposits on the Cinca and Gallego Rivers (NE Spain) based on OSL dating and soil stratigraphy. Global and Planetary Change. 2009;**67**:141-152. DOI: 10.1016/j.gloplacha.2009.01.001

[23] Turu V, Carrasco RM, Pedraza J, Ros X, Ruiz-Zapata B, Soriano-López JM, et al. Late glacial and post-glacial deposits of the Navamuño peatbog (Iberian central system): Chronology and paleoenvironmental implications. Quaternary International. 2018;**470**:82-95. DOI: 10.1016/j. quaint.2017.08.018

[24] Montserrat-Martí JM. Evolución glaciar y postglaciar del clima y la vegetación en la vertiente sur del Pirineo: Estudio palinológico. Zaragoza: Monografía 6 Instituto Pirenaico de Ecología; 1992. 147 p

[25] Brookfield ME, Martini IP. Facies architecture and sequence stratigraphy in glacially influenced basins: Basic problems and water-level/glacier input-point controls (with an example from the Quaternary of Ontario, Canada). Sedimentary Geology. 1999;**123**(3-4):183-197. DOI: 10.1016/S0037-0738(98)00088-8

[26] Carrasco RM, Turu V, Pedraza J, Muñoz-Martín A, Ros X, Sánchez-Vizcaíno J, et al. Near surface geophysical analysis of the Navamuño depression (sierra de Béjar, Iberian central system): Geometry, sedimentary infill and genetic implications of tectonic and glacial footprint. Geomorphology. 2018;**315**:1-16. DOI: 10.1016/j.geomorph.2018.05.003

[27] Mitchum RM Jr, Vail PR, Thompson S III. Seismic stratigraphy and global changes of sea-level, part 2: The depositional sequence as a basic unit for stratigraphic analysis. In: Payton CE, editor. Seismic Stratigraphy – Applications to Hydrocarbon Exploration. Vol. 26. Tulsa: American Association of Petroleum Geologists Memoir; 1977. pp. 53-62

[28] Mitchum RM Jr. Seismic stratigraphy and global changes of sea-level, part II: Glosary of therms used in seismic stratigraphy.

In: Payton CE, editor. Seismic Stratigraphy – Applications to Hydrocarbon Exploration. Vol. 26. Tulsa: American Association of Petroleum Geologists Memoir; 1977. pp. 205-212

[29] Vail PR, Hardenbol J, Tood RG. Jurassic unconformities, chronostratigraphy, and sea level changes from seismic stratigraphy and biostratigraphy. In: Schlee JS, editor. Interregional Unconformities and Hydrocarbon Accumulation. Vol. 36. Tulsa: American Association of Petroleum Geologists Memoir; 1984. pp. 129-144

[30] Van Wagoner JC, Posamentier HW, Mitchum RM Jr, Vail PR, Sarg JF, Loutit TS, Hardenbol J. An overview of the fundamentals of sequence stratigraphy and key definitions. In: Wilgus CK, Hastings BS, Kendal CGSC, Posamantier HW, Ross CA, Van Wagoner JC, editors. Vol. 42. SEPM Special Publications; 1988. pp. 39-45

[31] Embry AF. Sequence boundaries and sequence hierarchies; problems and proposals. In: Steel RJ, Felt VL, Johannessen EP, Mathieu C, editor. Sequence stratigraphy on the Northwest European Margin. Vol. 5. Norwegian Petroleum Society (NPF) Special Publications; 1995. pp. 1-11

[32] Catuneanu O. Principles of Squence Stratigraphy. Italy: Elsevier; 2006. 375 p. DOI: 10.1017/S0016756807003627

[33] Pous J, Ledo JJ, Marcuello A, Daignières M. Electrical resistivity model of the crust and upper mantle from a magnetotelluric survey through the Central Pyrenees. Geophysical Journal International. 1995;**121**:750-762. DOI: 10.1111/j.1365-246X.1995.tb06436.x

[34] Perkins AJ, Brennand TA. Refining the pattern and style of cordilleran ice sheet retreat: Palaeogeography,

evolution and implications of lateglacial ice-dammed lake systems on the southern Fraser plateau, British Columbia, Canada. Boreas. 2015;**44**(2):319-342. DOI: 10.1111/bor.12100

[35] Gilbert CK. Lake Bonneville. US Geological Survey Memoir. 1890;**1**:1-438

[36] Fitzsimons SJ. Sedimentology and depositional model for glaciolacustrine deposits in an ice-dammed tributary valley, western Tasmania, Australia. Sedimentology. 1992;**39**(3):393-410. DOI: 10.1111/j.1365-3091.1992.tb02124.x

[37] Fitzsimons S, Howarth J. Glaciolacustrine processes. In: Menzies J, van der Meer JJM, editors. Past Glacial Environments. 2nd ed. Elsevier; 2018. pp. 309-334. DOI: 10.1016/B978-0-08-100524-8.00009-9

[38] Mutti E. Turbidite systems and their relations to depositional sequences. In: Zuffa GC, editor. Provenance of Arenites. Dordrecht: NATO ASI Series C - Reidel Pub.(Springer); 1985. pp. 65-93. DOI: 10.1007/978-94-017-2809-6_4

[39] Bouma AH. Turbidites. In: Bouma AH, Brouwer A, editors. Developments in Sedimentology. Vol. 3. Amsterdam: Elsevier; 1964. pp. 247-256. DOI: 10.1016/S0070-4571(08)70967-1

[40] Berg OR. Seismic detection and evaluation of delta and turbidite sequences: Their application to exploration for the subtle trap. American Association of Petroleum Geologists Bulletin. 1982;**66**:1271-1288

[41] Bhattacharya JP, Walker RG. Deltas. In: Walker RG, James NP, editors. Facies Models: Response to Sea Level Changes. Vol. 1. St John's: Geological Association of Canada Geotext; 1992. pp. 157-178. DOI: 10.1002/gj.3350290317

[42] Vail PR, Audemard F, Bowman SA, Eisner PN, Pérez-Cruz C. The stratigraphic signatures of tectonics, eustasy and sedimentology, an overview. In: Einsele G, Riken W, Seilacher A, editors. Cycles and Events in Stratigraphy. Berlin: Springer-Verlag; 1991. pp. 617-659

[43] Milankovitch M. Kanon der Erdbestrahlung und seine Anwendung auf das Eiszeitenproblem. Belgrade: Zavod Nastavna Sredstva; 1941. 634 p

[44] Imbrie J. In: Imbrie KP, editor. Ice Ages: Solving the Mystery. London: Harvard University Press; 1986. 224 p

[45] Smith RO, Bryden HL, Stansfield K. Observations of new western Mediterranean deep water formation using Argo floats 2004-2006. Ocean Science. 2008;**4**(2):133-149. DOI: 10.5194/os-4-133-2008

[46] Florinet D, Schlüchter C. Alpine evidence for atmospheric circulation patterns in Europe during the last glacial maximum. Quaternary Research. 2000;**54**:295-308. DOI: 10.1006/qres.2000.2169

[47] Schott F, Leaman K. Observations with moored acoustic Doppler current profilers in the convection regime in the Gulf of lions. Journal of Physical Oceanography. 1991;**21**:558-574. DOI: 10.1175/1520-0485(1991)021<0558:OWMADC>2.0.CO;2

[48] Gilman C, Garrett C. Heat fux parametrizations for the mediterranean sea: The role of atmospheric aerosols and constraints from the water budget. Journal of Geophysical Research. 1994;**99**(C3):5119-5134. DOI: 10.1029/93JC03069

[49] Millot C. Heterogeneities of in- and out-flows in the Mediterranean Sea. Progress in Oceanography. 2014;**120**:254-278. DOI: 10.1016/j.pocean.2013.09.007

[50] Rohling EJ, Marino G, Grant KM. Mediterranean climate and oceanography, and the periodic development of anoxic events (sapropels). Earth-Science Reviews. 2015;**143**:62-97. DOI: 10.1016/j. earscirev.2015.01.008

[51] Rogerson M, Cacho I, Jimenez-Espejo F, Reguera MI, Sierro FJ, Martinez-Ruiz F, et al. A dynamic explanation for the origin of the western Mediterranean organic rich layers. Geochemistry, Geophysics, Geosystems. 2008;**9**:Q07U01. DOI: 10.1029/2007GC001936

[52] Rohling EJ. Review and new aspects concerning the formation of Mediterranean sapropels. Marine Geology. 1994;**122**:1-28. DOI: 10.1016/0025-3227(94)90202-X

[53] Calvet M. The quaternary glaciation of the Pyrenees. In: Ehlers J, Gibbard PL, editors. Quaternary Glaciations—Extent and Chronology: Part 1 Europe. Amsterdam: Elsevier; 2004. pp. 119-128. DOI: 10.1016/S1571-0866(04)80062-9

[54] Dansgaard W, Johnsen SJ, Clausen HB, Dahl-Jensen D, Gundestrup NS, Hammer CU, et al. Evidence for general instability of past climate from a 250-kyr ice-core record. Nature. 1993;**364**:218-220. DOI: 10.1038/364218a0